Vapor Tiger

I0405212

This is the full story of "global warming" also known as, "climate change;" the people, principles and forces driving it. It is the greatest fraud ever perpetrated. It was created by a man who for not being rewarded with the Presidency of the University of California, San Diego sought revenge on the world. It was created in anger.

After his work did real harm he had a fit of conscience late in life, but it had gone too far: The genies of avarice were out of the lamp. He died knowing well he set in motion an engine that could destroy the nation. We have very long worked to stop the process with education by making the science clear, but at a price. This is our final effort in this campaign. America is at a cliff edge on this matter.

The objective of this work is to teach you as much as there is to know how the atmosphere really works. The physics is simple, as well the math, but both take more explaining than can be done in a magazine article, television or radi o program and this book is the way to present the material completely and entirely. You will here learn as much as there is to know about "global warming."

Those familiar with math will find some of the material tediously over-explained, but we mean for everyone to understand it well. It is easy for the math sophisticated to skip over the step-by-step presentations, but not possible for those not so well prepared to understand what we are doing unless all the equations are thoroughly explained.

Table of Contents

The Beginning

The story of "global warming" begins with Roger Revelle, born in 1909 in Seattle, Washington. His family moved to southern California where he earned a BS in Geology at Pomona College in 1929. He went on to earn a Ph.D. in Oceanography at UC Berkeley, then went on to the Scripps Institute to teach and then serve in the US Navy as a staff oceanographer during World War II. After World War II he stayed at Scripps as a teacher and staff scientist.

San Diego in the 30's and 40's was a mix of military and old money retiring to a mild climate and easy living. The wealth was imported and held by people who were great builders by nature, training and experience during the war years. They impressed Roger Revelle, inspiring him to work for advancement on seeing what great wealth could mean. He felt science was overlooked and undervalued.

He was the Director of Scripps from 1950 to 1964 and a major force in getting the University of California to San Diego. Meanwhile he had wondered how increasing CO_2 was affecting the oceans. Atmospheric scientists noted it had been increasing since the 30's with the preparations for and fighting of WWII. The increase continued afterwards with the expansion of industry and the world rebuilding.

Dr. Revelle wrote a paper on the effects of increasing CO_2 in the air and seas. His conclusion should have been there would be no effect as the seas are huge and the amounts of CO_2 added are small, but Dr. Revelle postulated otherwise and got a big reaction! He was shocked with the interest and acclaim. He learned a controversial issue could work well for him. It was a turning point in his life.

Given his prominence and power at Scripps, as well as his role in bringing the University of California to San Diego, Roger Revelle expected to be named President of UCSD, but he was not. In a fit of pique he decided to move on with a new issue, population, as it had been in headlines thanks to books by Paul Erlich and other such alarmists.

Revelle put together an idea for a population study center selling it to a Harvard administrator for a new Center for Population Studies as his politics crystallized on the left. He resigned from Scripps, announced he was going to Harvard, the cutting edge of growing national socialism..

Population control is part and parcel of national socialism as they want to control everything for everyone. The free market, and free choice are anathematic to socialism.

Apparently, Revelle's Harvard connection was not able to get funding for a "population studies institute" as they had Dr. Revelle teach a science survey course. This had to be galling for him as it is something they would normally do to a new PhD, professor; not a man with a distinguished career expecting to direct his own institute! But it put him on the path of destiny as Albert Gore, Junior was one of his students.

While Director of Scripps, Revelle and a researcher wrote a paper linking carbon dioxide from burning fossil fuels to heating the atmosphere. That they did not publish it says very much about it. Where Roger Revelle was very well prepared in the basic sciences it would seem he knew well man's activities could not cause "global warming" as all the quantities and equations were well known. He must have realized that if anything more CO_2 in air would have a cooling effect by driving out the gas that was heating air, water vapor. All other gases are transparent to infrared, IR, sunlight heat waves, absorbing very little to none.

He reviewed the paper and had students check data on the effects of CO_2 in the atmosphere with an emphasis on that generated by the burning of fossil fuels. Albert Gore, Jr. participated in this work and it made a big impression on him. Unfortunately, not enough on Revelle to keep Albert Gore, Junior from getting a “D” grade in his only science course, but young Al Gore saw the political potentials.

Al's grades during his first two years were in the lowest quintile of the class, but he earned an "A" on his senior thesis, with the awkward title: "The Impact of Television on the Conduct of the Presidency, 1947-1969." He then graduated "AB Cum Laude" in June 1969 and was drafted for military service. However, everyone graduates "cum laude" from Harvard College.

As the son of a sitting US Senator, Albert Gore, Junior was not allowed to be in combat, but instead was sequestered in the basement of the Saigon Hilton Hotel under the guard of two MP’s as he worked on a book project.

The Revelle CO_2 paper was a great success and in 1974 Revelle was named President of the American Association for the Advancement of Science. Apparently the people there had seen the potential in political and funding power in the issue. Dr. Revelle was the cutting edge of what would become the greatest cash cow in the history of science, anthropogenic, man-caused, global warming! World War II had taught the political and defense suppliers the power of panic. There is great money to be made from fear.

Gore was so impressed with the Revelle project he based his 1992 book, "Earth in the Balance," on the proposition that also became the slide show "An Inconvenient Truth" earning Albert Gore, Junior an Oscar for something you would show to a General Science class, but not get half a Nobel Peace Prize. He also got $100 million from stock shares he was given to be on the boards of several Silicon Valley mega-companies, fees for the use of his name and participation in a carbon trading scheme that cost Goldman Sachs $200 million. They built a new electronic exchange building for trading that never happened.

Roger Revelle was a powerful man, a noteworthy scientist and a significant force in San Diego in the 1950s. There is no doubt he was largely responsible for the respect given Scripps Institute of Oceanography and for locating a new University of California campus in San Diego. There is also no doubt Roger Revelle made a major impact on Vice President Albert Gore, Junior's life. However…

In 1988 Roger Revelle admitted to serious doubts about carbon dioxide being a significant "greenhouse gas." He wrote letters to two Congressmen expressing his doubts. And, in 1991 co-authored a feature for the new science magazine "Cosmos" in which he expressed very strong doubts about CO_2 causing global warming while urging more research before any remedial actions were taken.

It is unfortunate Dr. Revelle did not make a good, well-explained physical case against CO_2, as he certainly could have, but we will here We have no doubt he could have quenched this issue with one well-placed book, which he certainly could have gotten published with a phone call.

On hearing of his doubts Mr. Gore said Revelle was senile publicly, but refused to debate him on global warming. To this day he refuses to debate anyone. Many offers of tens of thousands of Dollars have been made for such a debate, but Albert Gore, Jr. is so insecure he will not allow media questions at any event where he speaks and he will not "Q & A" with audiences. The simple fact of the matter is he knows nothing about science, but a lot about money.

In 1991 Roger Revelle made a speech at the very private summer enclave of powerful men and politicians at the Bohemian Grove in Northern California. He apologized that his ideas had sent so many in the wrong direction on global warming. Some there were shocked, but all were impressed with the clarity of his talk and arguments in spite of Albert Gore, Jr's diagnosis of his senility. It is tragic no recording or transcript of that speech now exist.

Mr. Donald Michael Schmidtman of the San Francisco Bay area was there and remembers the Revelle speech. He has talked about it in some detail. Other people who were there have confirmed Schmidtman's recollections.

The final irony is that Albert Gore, Junior received the first Roger Revelle Science Award, an honor named after the man who set Albert Gore, Junior on his global warming campaign, published book, slide show Oscar, half a Nobel Prize, millions of Dollars and all of it on a "D" grade in a "dumbbell" science survey course.

Roger Revelle realized his original concept about CO_2 was wrong and in spite of all the money it brought to scientists he could see that it was a great danger to America, but he could not put the genies back in the lamps. Revelle died of a failing heart in 1991. Or was it a broken heart?

The Atmosphere

The atmosphere is a mixture of gases: 78% nitrogen, 18% to 20% oxygen, one to four percent water vapor and one percent "trace gases," of which there are nine.

A "trace gas" has less than one percent of the atmosphere and is deemed physically insignificant: Argon is 0.937%, CO_2 is only 0.039%, followed by seven gases with even less quantity and significance, notable only because they are there and we have developed ways of identifying them.

The most important fact about trace gases is that we have perfected our analytical methods well enough to find them and determine their quantities. That is their significance.

There is so little carbon dioxide in the atmosphere we did not know how much, with accuracy, until well into the 20th century. The chemical methods for determining how much CO_2 is present were unreliable so we were not sure of the quantity until instrumental methods were invented.

Even today CO_2 standards are based on lab gas analysis, the most unreliable area in chemistry, but they are deemed valid only because of many trials getting consistent results. Nonetheless, scientists are always haunted by the chance of only having made the same mistakes consistently.

We knew carbon dioxide had to be present in air as it is a product of combustion, decay of limestone, dead matter, both plant and animal. Every plant uses carbon molecules from air or soil where dead plants and animals decay, but soil sourcing of CO_2 is overlooked by agriculture. Soil CO_2 is a major source of the gas; the reason for “plowing under” and “potting soils,” a practice of putting decaying organic matter in the ground. That CO_2 is the reason for doing this is never noted in the literature!

CO_2 is not very soluble in water, which only dissolves 1.6 grams per liter at 20 Celsius degrees. Where green plants have to capture most of the carbon that constitutes 42% of their dry body mass from air they have to transpire great quantities of water to capture the amount of gas they need.

1680 liters of water are needed for every pound of dry plant mass. Usually more, as the temperature of leaves in sunlight approaches 30 Celsius degrees where the CO_2 solubility is 1.25 g/l, 22% less than at 20 Celsius degrees. Reducing CO_2 in air only makes the shortage worse.

For 150 years greenhouse owners have burned charcoal and propane in closed greenhouses to elevate the CO_2 concentration to levels on the order of 1,000 parts per million and greatly enhanced growing products without deleterious effects to the workers.

CO_2 does not affect animal life until it exceeds 15,000 parts per million, 1.5% CO_2, and green plants would love it, but that amount is our signal to inhale. We would find it difficult to live in such air. We would be gasping all the time.

The bulk of atmospheric CO_2 is produced by decaying limestone dissolving into bicarbonate ions in fresh water going to the seas where mollusks and algae build their bodies of shell and cellulose. A little CO_2 escapes regardless it is10% more soluble in sea water than fresh water. Sea water is basic, having a pH of 8.2, and CO_2 combines with sodium and calcium ions in sea water to form bicarbonate ions. 97% of all CO_2 in the air is from limestone. It is natural.

Our atmosphere has the configuration of a 1/3rd inch soap bubble. The thin soap skin models our atmosphere. We live in that narrow space on a very large rock in space. Above it there is nothing. The talk of "shields" given by ozone and CO_2 is mythical. The physical nature of the atmosphere is such that these things are not possible.

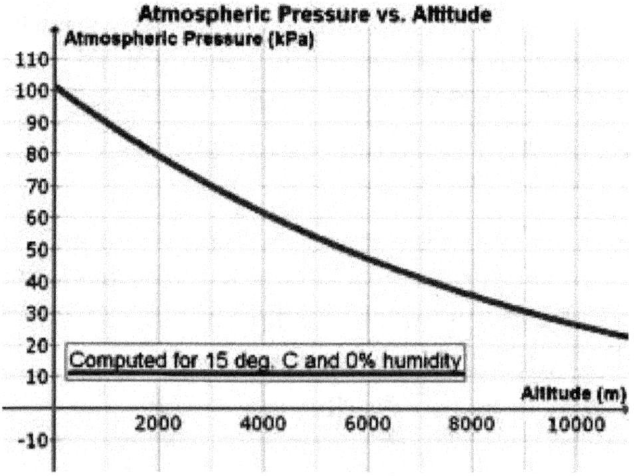

Air pressure declines with altitude in a smooth curve that is almost a straight line showing 80% of the atmosphere is below 33,000 feet or 10,000 meters. We can survive with no less than 70% of sea level pressure and are limited to living between sea level and 2,400 meters or 8,000 feet. Our space is a thin skin on an 8000 mile diameter rock.

The amount of water vapor declines faster with a rise in

altitude than pressure alone requires as it is temperature correlated: It varies directly with temperature and at the zero degree Celsius temperature of the upper troposphere is virtually absent. This is the case at 6000 meters or 19,685 feet. This is far above where any animal can live and virtually no atmospheric heating occurs despite the fact CO_2 is the only IR absorber present, but the very low quantity, 0.04% of sea level, is further diminished by the fact CO_2 is a heavy molecule that diffuses less well than that the other, lighter, molecules in air. Where it is 0.04% at sea level it is only 0.016% at 6,000 meters. Claims that CO_2 reflect or re-radiate heat to the surface are fraudulent misrepresentations of basic atmospheric science.

At the surface there is 14.7 pounds per square inch of air pressure the per square foot pressure but only 2.12 pounds on each square foot at an altitude of 20 miles. The surface pressure is so close to a metric ton we can use one ton per square foot in calculations.

Earth is a sphere and the surface area of spheres is found by "4 pi r^2" where "r" is the radius and the product is: 201,000,000 square miles for the surface area of Earth.

Scientists have to deal with many large numbers and use a system called "Scientific Notation" that simplifies this to 2.01 x 10^8. It is much easier to use in computations as all operations are with small numbers.

Where 10^1 = 10, 10^2 = 100, 10^3 = 1,000 we say that 201,000,000 is "2.01 X 10^8th." For Scientific Notation the rule is to express the number as "unit, decimal point, two figures and ten times the appropriate power of ten" which we find by counting the digits. Each digit is a factor of ten.

Each square mile is 2.78 x 10^7 square feet and has that many tons of air pressing on it. 2.78 x 10^7 is 27.8 million and we can calculate the mass of the atmosphere in tons by multiplying the area of Earth in square miles by the mass of air on every square mile with:

2.01 x 10^8 mi^2 x 2.78 X 10^7 ton/mi^2 = mass of atm.

We multiply the digits first: 2.01 x 2.78 = 5.59 and then add the exponents to multiply them. Just as 10^2 x10^2, or 100 x 100 is 10,000, 10^2 x 10^2 is 10^4 so 10^8 X 10^7 is 10^15 and we have 5.59 x 10^15 tons of air on Earth.

5.59 x 10^15 tons of air is a lot of air spread over a very large sphere.

This calculation says the mass of the atmosphere in metric tons is 2,117 is within 96% of the 2,200 pounds per metric ton. If we must be totally accurate in conventional terms we will reduce it by 4% to 5.37 x 10^15 metric tons.

Of that 5.37 x 10^15 metric tons 4.19 x 10^15 tons is nitrogen, 9.97 x 10^14 tons is oxygen, 1.66 x 10^14 tons is water vapor, 5 x 10^13 tons is argon and 2 x 10^12 tons is CO_2. Generally, in statistics and physical science we have to see more than 5% change in an entity to be significant.

Argon and carbon dioxide are seen by physical scientists as insignificant, not worthy of consideration in the physics of the atmosphere, and many scientists have objected to all the wild claims regarding CO_2, but panic pushers seeking power and new taxes drowned them out and with major media seeking to sell more newspapers, books and build bigger TV audiences by frightening people to think man could destroy the planet. Thus, the lie has persisted.

It is flattering to some that man could destroy Earth, but as we will show, that is simply not the case. Just as a family needs a father it can trust America must have a President and government it can trust. All the systems are in place to give us the truth of any matter, but have been corrupted by ambitions for money and power, in this case, by taxing the source of 80% of all our energy.

Nitrogen and oxygen are the bulk of air, but water vapor is significant in three very important ways that cannot be overlooked in any discussion of the atmosphere. However, anthropogenic global warming people ignore it totally.

Clouds

Water vapor condenses to tiny water drops forming clouds that are very reflective and can cut sunlight by 80%. But, usually the reduction is no more than 50% due to the space between the drops. Rain clouds are the way nature moves water around the planet. Rain is pure water, supporting life and required by everything living on land.

Atmospheric Heating

Water vapor is far and away the principle "greenhouse gas" and responsible for 99.8% of all atmospheric heating. There is no other gas in air that comes anywhere near the water molecule in terms of capturing infrared, heat waves, from sunlight. CO2 does one seventh as well and methane not even one percent as well. This is easily seen in official absorption charts at the American Meteorological Society.

The analysis of infrared ray energy absorption is somewhat complicated because wave energy is inversely proportional to wavelength. When we compare IR absorbing gases they may have similar areas of energy absorption, but capture greatly different amounts of energy due to where they do it in the IR spectrum. Energy is inversely proportional to the square of the wavelength.

CO_2 has a smaller area of IR absorption and most of it is in the low energy side of the spectrum where water vapor has more and in the higher-energy, shorter-wave portion of the spectrum. We will do an elaborate analysis and summation of these spectra in order to fully explain this phenomenon.

All of this is important as the atmosphere is a physical system and to really understand it you have to deal with complicated concepts that have not previously been shown in a way most people can understand. Those who believe "More money for science..." is a noble objective have sold their souls.

Greenhouse Gas

The "climate change" people make a telling error. Earth is not a "greenhouse." They are covered with clear panes that trap heat energy, infrared radiation, during the six hours it can enter such a building and trap it. Molecules first absorb and later re-emit but at angles that cannot escape the enclosure. Thus, temperature in the glazed space rises.

Earth is not such a structure. Heat energy that is absorbed can leave the atmosphere easily and much of it does. The "CO_2 greenhouse effect" is a bad analogy, myth and lie, but there is one phenomenon the "greenhousers" may offer in defense:

Mirages are caused when the atmosphere layers due to a difference in temperatures in layers. This results in an effect which if struck by light at a very low angle, literally a few degrees, the boundary functions reflectively in the manner of a lake on which stones may be "skipped" if thrown at an angle parallel to the surface and water acts as a solid. This will not make a "greenhouse."

The American people were defrauded of $322 billion from 1988 to 2000 by scientists who believed they were entitled to such wealth because they are smarter than ordinary folks and spent more time in school. However, kissing butts and plagiarizing publications is not valuable. They have ruined science education and will soon ruin America. The mantra is: "Greenhouse gas" and therein lies a very big mistake:

Greenhouses are seen on estates or farms for growing out-of-season or exotic flowers, fruits and vegetables. They create warm micro-climates in cold ones and now have increased CO_2 plants need to make stems, stalks, leaves, flowers and fruits. The "greens" tell us such CO_2 levels would destroy the atmosphere, while they would not, but that is not the only problem.

The classic "greenhouse" is covered with glass panes in order to capture sunlight so the heat generated by it is trapped to make a microclimate. The physics of this is often not understood and it is no longer taught in schools. Science education has been distorted for the sake of getting

Federal grant money by scaring the people with a species - flattering myth that man can destroy the planet. We could not do that if we wanted to and the great weapon of the fear mongers is the fraudulent "greenhouse" analogy.

An analogy is a simple thing often used to explain a more complex thing and in science it is usually something small, like atoms, that require such teaching devices. We often use small balls to represent protons, neutrons and electrons and they are very likely not like that at all, but the analogy works in our minds. We are relating the world we can hold in our hands to a dimension so small we will never see it, but in any case an analogy has to be valid: "Greenhouse" is not.

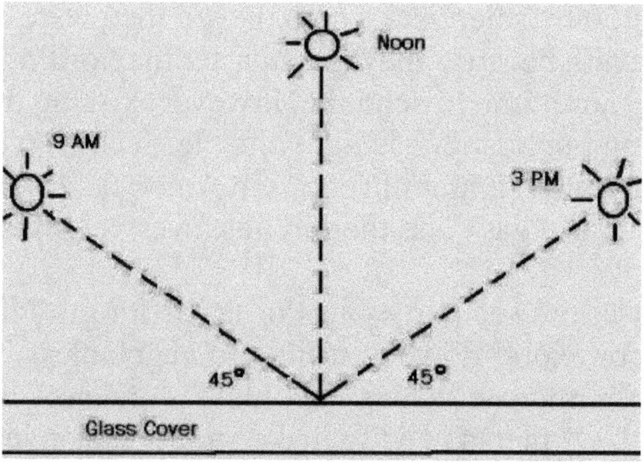

At any angle below 45 degrees an incident solar beam is totally reflected and nothing goes into the glass enclosure. It is like a rock skipping on a pond. This is a surface phenomenon is seen on any liquid or polished solid and a property of transparent materials. At angles above 45 degrees an increasing portion of the light passes through the glass. That portion is equal to the sine of the angle which for 45 degrees is 0.707 and for 90 degrees it is 1.00

or as percentages 70.7% and 100% respectively with a mean, or average, value of 0.85, or 85%.

Solar panels and greenhouses are about 85% efficient in capturing energy for six hours of a day if the glass contains no iron ions as iron absorbs some of the spectrum. Glass with iron in it looks green on edge and that without iron appears blue or clear so iron can be easily avoided.

Solar radiation entering a greenhouse is absorbed by plants and objects in the enclosure, warming the air for a true "greenhouse effect." The atmosphere does not do this.

"Climate Change" promoters claim stratospheric CO_2 traps solar infrared, IR, heat energy, holding it in the atmosphere by reflecting it back to Earth. Gases do not form "shields," as they claim. Reflecting surfaces can only be formed by solids and liquids where molecules are in contact. In gases they are not in contact.

CO_2 is a "trace gas" in air with a higher concentration in lower levels than the stratosphere due to a high molecular weight of CO_2. Nitrogen has a molecular mass of 28 awu and oxygen 32 awu. The mean mass is 29 awu. Thus a prototype "air molecule" has a molecular mass of 29 awu and CO_2 has molecular mass of 44 awu.

Where CO_2 is generated on the surface, per Graham's Law, and gases diffuse at rates inversely to the square roots of their densities which are 0.80 grams/liter for H_2O and 1.40 g/l for CO_2. We should expect to see water vapor diffusing at [1.40/0.894] = 1.56 times the rate of CO_2 and there will be greater amounts of it up to about 5,000 feet where low temperatures cause water to precipitate to form clouds.

The CO_2 concentration rises relatively, but is still so small the difference is insignificant as seem by low temperatures at higher altitudes. If "global warmers" were correct the upper atmosphere would be hotter, not colder.

The atmosphere cannot act like a "greenhouse" in any case and trap IR energy. There are no "greenhouse.gases" and "man-caused global warming" is false. It is not physically possible for the atmosphere to be a greenhouse.

John Tyndall

Quantitative work with the atmosphere began with English physicist, John Tyndall (1820-1893). He wrote of gases "radiative properties," that we now call "absorption." He did his work in 1857. He worked with a "spectrograph" that uses a prism to spread light waves into a multicolored "rainbow" pattern seen in physics texts and raining skies.

Photography had been invented 18 years before Tyndall did his work. He was among the first to use photographic film in physics realizing it could be used for quantitative work in the visible spectrum as well as beyond either side, of red or violet as silver was deposited in areas beyond his ability to see them. He knew that radiation must be there.

The invisible rays beyond red had low frequencies with longer wave-lengths and energies so they were "infrared," or below "red." The invisible rays beyond violet were of higher frequency and energy as well as being shorter in wavelength so they were called "ultraviolet," for above "violet."

These invisible waves are recorded by photographic film. The greater the quantity the darker the deposit. When the light was spread with a prism physicists could measure the spectrum directly on film, determine densities accurately with comparison system using a “standard candle.”

In the 19th century physicists defined a “Standard Candle” which was made with specified wax, in a certain diameter with a standard wick such that everyone could make such “standard candles” to produce the same quantity of light.

Light intensity varies with the square of the distance. The illumination of a translucent oiled paper at one foot would be four times greater than that at two feet, for example, and by illuminating a developed film by one standard candle, compared to a sheet illuminated by a movable candle they could determine relative optical density of the film with the movable candle to produce a match and compute the illumination and density in percentage, or the equivalent decimal value, after the apparatus had been calibrated.

Tyndall used the spectrophotometer with photographic film to measure the energy absorption of atmospheric gases between 0.1 and 16 microns, the wavelengths of “heat” waves. A micron is a millionth of a meter.

Tyndall found nitrogen and oxygen were transparent to heat waves, IR, infrared, long wavelength radiation and absorbed none of it. However water vapor, the third most abundant gas in air, absorbed IR, heat waves, very well, far better than any other gas in air! From the NASA "Earth Observatory" website "On the Shoulders of Giants – John Tyndall" (paragraph three, sentence two) wrote: "He concluded that among the constituents of the atmosphere, water vapor is the strongest absorber of radiant heat and is therefore the most important gas in controlling Earth’s surface temperature."

NASA removed this sentence in 2011 and changed the piece to support anthropogenic global warming when in truth Tyndall did not consider CO2 in his work. But, that is not all NASA did to John Tyndall. He was one of the first to record and display data in charts. NASA gave him no attribution and worse: They actually printed his charts upside down to make it appear that CO2 absorbs infrared, IR, radiation much better than it does!

Tricks of this kind have gotten NASA billions of Dollars in public monies for projects that are nothing more than full-employment programs for science Ph.D.s.

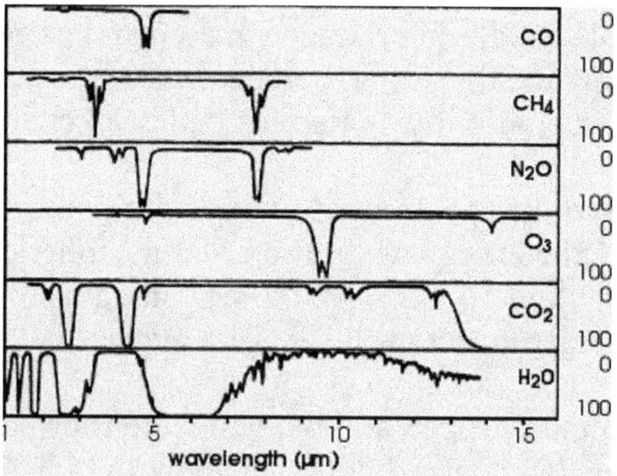

On first glance it appears CO2 is a much better IR, infrared absorber than water vapor as absorption is measured by the area under the curves! Then we noticed the chart is upside down! Normally zero, "0," is in the lower left corner and the maximum, "100," in percentage, would be in the upper left corner.

And, the charts are stacked such that "100" and "0" figures

are displayed in a manner to confuse, we are quite certain, given the upside-down way in which the data are printed. We corrected the chart and it is quickly apparent CO_2 is a very poor absorber of IR radiation from sunlight and this is confirmed by every other published chart!

We inverted the chart, removed the original labels and numbers, ask you understand “0%” and “100%” indices are then where they should be, in lower and upper left corners respectively.

That this was ever done by a NASA employee is shocking and it should be prosecuted as a felony with prison time.

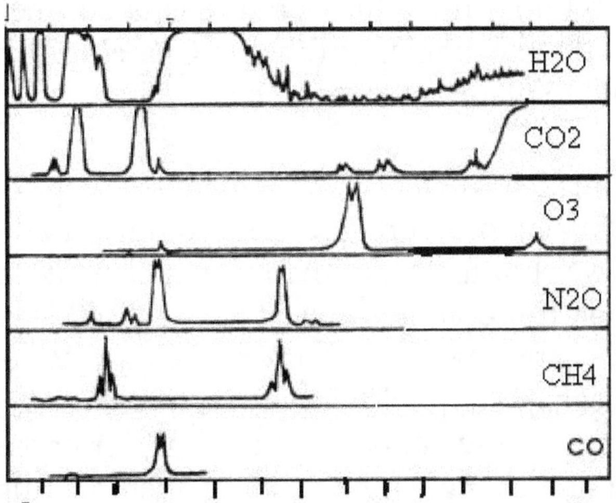

Note that methane, CH_4, is a very poor absorber of IR as the areas under the curves are miniscule in spite of Ja mes Hansen and Heidi Cullen’s claims it is “500 times the greenhouse gas as CO_2!” What really makes this claim utterly ridiculous is that methane rarely has more than 18 parts per million, 0.0018% and it auto-oxidizes in sunlight, even at twilight where it is known as "wil-o-the-wisp."

What NASA has done to John Tyndall is a crime against science and this was not their last such felony.

Vapor is gas that can be liquid at ambient temperatures. Water is liquid at any temperature between zero and 100° Celsius degrees at one atmosphere, but molecular motion caroms molecules out to become gas at any temperature in that range. The activity is greater at higher temperatures.

Once a gas, water remains so until pressure increases or temperature falls enough to precipitate it as water. If the vapors are given higher pressure or lower temperature some revert to liquid if the temperature is under the boiling point. This is the mechanism of the Le Chatelier Principle as it is applied to water and water vapor.

At temperatures less than 100° Celsius water vapor is forced out of air by pressure, reduced volume or adding a gas with a lower boiling point than water. This is seen as aircraft temporarily compress air in passing and leave trails of condensed water that fades back to the gas. When a gas cylinder is discharged quickly or an explosion makes a shock wave we also see condensations. Water or ice form depending on the change in pressure and temperature.

Water vapor is limited to 4% by the gas/vapor equilibrium in our normal range of pressures and temperatures. If the temperature declines the vapor precipitates to form liquid: clouds, fog, mist, rain, sleet or snow. If we add more of a true gas to saturated air water is forced from air as it acts as it were in a sealed flask; it is the bottle of gravity.

Only a small fraction of water molecules have enough energy to escape liquid water and be vapor below 100° Celsius. In a cubic foot of air at 20 Celsius degrees, (68 F) the maximum amount of water as vapor is 18 drops, 0.9 milliliter or cubic centimeter.

Our atmosphere is open to space, but effectively is in a "gravity bottle." Four billion years ago Earth had two times as much air as it has today. In another four billion years 99.99% of it will be gone. Earth will be like Mars today. Man will have been gone long before that happens, certainly in one billion years as we cannot survive with much less air than we now have, but other forces of nature will have taken us out long before that.

For this work we treat the atmosphere as if it were a sealed flask as the rate of change is insignificant to us given our scale of time. That our atmosphere is a contained system permits us to use the laws of equilibrium to work with it.

Physical and chemical equilibria are like a "teeter-totter." When one side rises the other falls. The "atmosphere as bottle" concept supports the equilibrium system concept.

The boiling point of water is almost 300 Celsius degrees up the scale from those of nitrogen at -196° Celsius or oxygen at -183° Celsius. Liquid nitrogen and oxygen have much greater tendencies to be gas than does water.

Rain, dew, mist, clouds, sweat and frost are products of adjustment in the gas equilibrium between the nitrogen, oxygen and water vapor which are 99% of air. Of the nine "trace gases" making one percent of air, carbon dioxide is the second largest fraction, but at 0.039% is insignificant.

Tyndall worked with air containing 3% to 4% water vapor in England. It varies around the world between one and four percent, but for all areas over the seas, forests, fields and jungles it is about three percent. In the driest areas of Earth water vapor approaches zero, but even there it has 26 times as many molecules as carbon dioxide. These areas are relatively small, insignificant and without weather.

Nonetheless, John Tyndall analyzed carbon dioxide and he found it also absorbs IR, but much less than water vapor. When the energy and wavelength relationship is included the difference is about seven to one for water vapor over carbon dioxide in terms of energy absorption, molecule for molecule, as we will later demonstrate thoroughly.

In Tyndall's time carbon dioxide had 280 parts/million or 0.028% of the atmosphere. Carbon dioxide molecules are 2.44 times as heavy as water molecules. With all factors considered, water vapor is 560 to 1,200 times as significant in absorbing IR as carbon dioxide for most of the planet; all of that over water and much of the land. Water vapor is thus responsible for up to 99.9% of all IR captured, but water vapor is of no interest to governments because they cannot find ways to control and tax it.

Svante Arrhenius

Late in the 19th century Svante Arrhenius, (1859-1927) the greatest Swedish chemist in history, author of the Proton Donor theory of acids, investigated IR capture seeking a way to warm Sweden. It is a cold, dark country much of the year, but fossil evidence revealed bananas grew there as recently as 6,000 BC. This greatly excited Arrhenius.

Arrhenius studied the work of Tyndall and thought where CO_2 absorbed IR he could warm the Swedish climate by burning huge quantities of coal to first heat the air and also increase IR capture with CO_2. This was, of course, before Le Chatelier. In 1897 Arrehenius wrote a paper on his idea and published it in the leading physics journal of the time, "Physique," of Heidelberg, Germany. The paper was little more than a curiosity. Two years later he wrote another paper recanting the first after having done more work and finding his original concept invalid.

Today the pro-global warming people cite the first paper and ignore the second. For a decade the NASA website, "On the Shoulders of Giants" had a translation of the second paper on the site as it was Arrehenius' conclusion, but they removed it in 2007 and now only cite the first.

In 1905 Arrhenius wrote "Worlds in the Making" for the general public and described a "hot-house theory "of the atmosphere, saying Earth's temperature is about 30 degrees warmer than it would be due to "heat-protection action of gases contained in the atmosphere," clearly attributing most of it to water vapor, but that fact is now omitted by those using Arrhenius name to promote taxing carbon.

Arrhenius' book "Worlds in the Making" includes the idea of an increase in carbon dioxide to the atmosphere by burning fossil fuels would be beneficial to make Earth's climates "more equable," stimulating plant growth, and providing more food for a larger population by adding more CO_2, but that is never noted by anti-fossil fuel people as they continue to pick-and-choose from his writings in a way that insults his reputation. There appears to be no limit to the crimes against science and the great men of science the "global warming gang" will not commit. It is all in their quest for more money and power.

Le Chatelier and the Atmosphere

Henrie Le Chatelier (1850 – 1936) was one of the great physical scientists of all time, best known for his concepts in chemical and physical reactions. He declared every reaction is an equilibrium system.

If you have ever seen a CO_2 fire extinguisher in action you saw a great cloud of white mist appear to come from it. That mist consists of some CO_2 freezing as it expands, but much of the white mist is water vapor driven from air by CO_2. Why should this be?

The Le Chatelier Demonstrator

We can demonstrate how the addition of CO_2 works with an easily assembled demonstrator from a polycarbo nate plastic soda bottle into which two cups, half a liter, of water and one level teaspoon, 8.5 grams, of sodium bicarbonate are dissolved.

The perforated plastic top is fitted with a "medicine dropper" we have inserted with the aid of a Phillips screwdriver. The head is heated by a candle and the point pushed into top to melt the plastic and make a hole for the dropper. It is then filled with white vinegar, five percent acetic acid, CH_3COOH. When the acid is dropped into the sodium bicarbonate solution CO_2 gas is generated and seen as small bubbles. They rise to the surface and upset the equilibrium of the saturated air to make a very fine mist. If you have difficulty seeing it you can see it in a darkened room with a small flashlight.

Casting the beam of a small flashlight across the top of the water surface will show the formation of tiny droplets of water. They are so small they will not fall, but form a small cloud in the bottle. This happens to maintain the physical equilibrium of the situation.

$$[H2Og] \times [CO2g] / [H2O] = Kt$$

The water and carbon dioxide in the numerator are gases, thus the "g" subscript. Water in the denominator is liquid.

Adding CO_2 means H_2O gas will decrease as it can change state at room temperature so Kt, the constant at temperature "t," does not change thus keeping the integrity of the system. (Brackets denote "moles/liter.")

Where water vapor molecules are seven times better at capturing infrared (heat) energy, IR, from sun light as CO_2 this cools the air by reducing the amount of energy captured from light.

The great French chemist and educator, Henri Le Chatelier first wrote of this in 1925. His concepts explain the period between 1930 and 1970 when our atmosphere cooled much to the consternation of anthropogenic global warming panic pushers who have been caught changing this data.

During the period from 1930 into 1970's we increased the amount of CO_2 as we prepared for, fought World War II and rebuilt the world. It was only Jimmy Carter's ruin the world economy that brought the temperature decline to an end.

According to Jim Hansen the atmosphere should have gotten hotter, but it did not and it was during this period he lost all his hair. Now it appears Barack Obama has had the same effect as Jimmy Carter as we are now in a cooler, drier period. Thus, if we want to prevent an oncoming warm cycle we must burn more carbon based fuels, not less.

The Vostok core studies say we are 40,000 years overdue for a major cooling period and CO_2 production will only increase that effect so Dr. Hansen put his money on the wrong horse. Where the truth is so obvious will he be able to admit his error and speak against carbon taxes for global warming? We doubt it.

Given this truth do we now champion carbon taxes? No because CO_2 is a trace gas in air and utterly insignificant by definition, making but 0.2% of atmospheric heating. It is totally insignificant. Carbon combustion makes 80% of all our energy, only 3.22% of the annual total, natural production and we should continue to use it for our benefit as well the green plants that feed us.

Svante Arrenheis, Sweden's greatest chemist knew this in the 19th century and published it in the leading German physics journal, but Green Party people only cite his 1897 book speculating CO_2 could be used to heat the air of Sweden. After more work he published an 1899 book rescinding his earlier ideas, but "global warming" partisans refer only to his 1897 publication.

When seen as an equilibrium system the atmosphere has only one component that can leave it by changing state: water vapor changes to tiny droplets of liquid. No other gas can as our, ambient temperature, is too high. Nitrogen and oxygen are gas above, -196°C or -183°C, respectively.

The Le Chatelier Principle states: "The product of the concentrations of the reactants over the product of the concentrations of the endproducts equals a constant which for the reaction A + B = C + D may be expressed as:

$$\frac{A \times B}{C \times D} = Kt$$

While this expression gives us the constant it is not a usable equation in terms of products. If we wonder about the amount of "D" to be produced we will divide both sides by "A" and "B," as well as multiply both sides by "C", following the rules of algebra, to produce:

$$D = \frac{Kt \times C}{A \times B}$$

Where the atmosphere has only one substance in it that can change state in the range of temperatures we find on Earth we can use this to predict what will happen when we change the composition of the atmosphere by adding CO_2.

This is an exercise that "global warming" promoters never perform in their papers and it will soon become obvious why that is the case. We make special note of it as it is the "crux of the matter," as they said in old English movies.

The atmosphere has 78% nitrogen, 18% oxygen and three percent water vapor plus one percent "trace gases," that are normally ignored. The inert gas argon has 0.9% and a list of increasingly rare gases including CO_2 with only 0.039% and methane with 0.00018%. Methane is insignificant as it auto-oxidizes in sunlight to make CO_2 and water vapor.

Evaporation and condensation are the significant physical reactions in our atmosphere. "Kt" is valid for a particular temperature. We can predict "Kt" values by finding a few, putting them on graph paper and then draw a curve to find values for higher and lower temperatures. The expression atmospheric physicists use for the atmosphere would not include CO_2 because it is a trace gas. It would read:

$$\frac{[N_2] \times [O_2] \times [H_2O \text{ gas}]}{[H_2O \text{ liq.}]} = Kt$$

Again, the Le Chatelier Principle says, "The product of the concentrations of the reactants over the product of the concentrations of the products equals the constant, "K," at temperature "t," i.e. "Kt."

In the atmosphere the only substance that can change state is water vapor by going to liquid. One liter of air has a mass of 1.29 grams: 1.01 g is nitrogen, 0.232 grams is oxygen, 0.0387 grams is water vapor and 0.000490 is CO_2.

A "mole" is the molecular weight in "atomic weight units" converted to grams to have equal numbers of atoms and molecules in a system. To determine the number of moles of nitrogen in air we divided the 1.01 grams in a liter by 28 as that is the sum of the 14 atomic weight units, awu, for each of the two atoms of nitrogen per molecule. Same way for oxygen's 0.232g by 32 awu, from two at 16 awu and water's 0.0387 by 18 from two hydrogens, at 1 each and one oxygen at 16 atomic weight units.

A liter of water, 1,000 grams, contains 55.5 moles by 1,000g/18 g/mole. Expressed as moles the figures for one liter of air are 0.0361 mole of nitrogen, 0.00725 mole of oxygen and 0.000215 mole of water vapor. In this work all quantities are in moles per liter.

When we put the numbers in an expression based on the Le Chatelier Principle we find that K is equal to:

$$\frac{0.0361 \times 0.00725 \times 0.000215}{55.5} = 1.01 \times 10^{-9}$$

Scientists write very large and very small numbers in "Scientific Notation." The form is "unit, decimal point, two figures, times ten to an exponent."

The constant "K" here is 0.00101 in decimal notation that is inconvenient in math computations where in scientific notation we multiply, divide or add the digits and powers separately to re-assemble it as 1.01 x 10^-3.

CO_2 is omitted from this expression because the physical scientists have long agreed there is so little CO_2 in air it is of no consequence. To document that fact we will write an expression using CO_2 and recalculate the system.

The number of moles per liter of CO_2 in air is 1.74 x 10^-5 mole. Adding this to the numerator does virtually nothing to the situation. To demonstrate that and why we say "CO_2 caused global warming is nonsense," we will revise the equation, solving for water vapor, the cause of atmospheric heating. The expression is originally:

$$\frac{[N_2] \times [O_2] \times [H_2O \text{ gas}]}{[H_2O \text{ liq.}]} = Kt$$

If we divide through both sides of the equality sign by "N2," "O2"and "H2O gas," then multiply both sides by "H2O liquid" and unit cancel:

$$\frac{[\cancel{N_2}] \times [\cancel{O_2}] \times [H_2O\ gas] \times [\cancel{H_2O\ liq}]}{[\cancel{N_2}] \times [\cancel{O_2}] \times [\cancel{H_2O\ liq}]} = \frac{Kt \times [H_2O\ liq]}{[N_2] \times [O_2]}$$

Cleaning it gives:

$$[H_2O\ gas] = \frac{Kt \times [H_2O\ liq]}{[N_2] \times [O_2]}$$

To check the validity of our work we will put in the molar numbers to test whether or not the "model," is correct.

$$2.12 \times 10^{-3} = \frac{1.00 \times 10^{-8} \times 55.5}{0.0361 \times 0.00725}$$

And the figures check as the quantity of water vapor here is within the rounding errors for what found earlier, so we continue by adding the CO_2 quantity to the numerator and recalculate the equilibrium constant for that case.

$$\frac{0.0361 \times 0.00725 \times 0.00212 \times 1.74 \times 10^{-5}}{55.5} = 1.74 \times 10^{-13}$$

This is the constant, "Kt," to model the atmosphere with CO_2. The question then becomes, "What happens if we increase the CO_2 in the system?" For that we return to our operating expression to calculate the change in the amount of water vapor the system as increased CO_2 will put some out of the system as clouds, rain, hail, sleet or snow. It is all in the equation:

$$[H_2O\ gas] = \frac{Kt \times [H_2O\ liq]}{[N_2] \times [O_2] \times [CO_2]}$$

And, inputting the numbers yields:

$$\frac{1.76 \times 10^{-13} \times 55.5}{0.0361 \times 0.00725 \times 1.74 \times 10^{-5}} = 2.14 \times 10^{-3}$$

The equation, or model, shows the presence of CO_2 makes the very slight difference of 0.02×10^{-3} moles of H_2O in the air. Suppose we double the CO_2 to 2.18×10^{-5} moles or 780 ppm which is a figure that will cause "global which is a figure said calamitous by global warmers

Doubling the quantity, in moles, of CO_2 to 2.18×10^{-5} will reduce the number of moles of water vapor to 6.75×10^{-6} moles per liter! This would have a chilling effect on the atmosphere because water vapor is seven times the absorber of IR, heat energy, from sunlight as carbon dioxide, molecule for molecule. Fortunately this is not possible.

The atmosphere has a mass of 5.59×10^{15} metric tons so at 0.04% the amount of CO_2 is 2.24×10^{12} metric tons and to double it we would have to generate much more as it dissolves in the seas, then precipitates with calcium ions as carbonate to make mollusk shells and fish bones or fall to the bottom as limestone. The oceans are an infinite sink for CO_2, but to demonstrate let us turn this off and proceed.

We will generate 2.24×10^{12}, or 2.24 trillion tons while the CO_2 made by nature and man equals 186 gigatons, thus we make 1/12th! Of that man makes only six gigatons, 1/373rd of 2.24×10^{12} tons. Therefore, if man doubled his use of fossil fuels he would need 373 years to complete this task and only if the seas stopped dissolving CO_2, as they do every day and green plants would have to stop consuming CO_2, as they do every day. And then, the temperature would fall and not rise as the international socialist global warmers fear!

While this should long ago have been the end of the story as Ph.D. after Ph.D. did not bother to do the math we have here, it should be obvious why this is the case if you add up all the grants these people have gotten for billions of Dollars to keep the myth going and ruin our economy in the process. This should be a felony and all degrees of the perpetrators be ripped up and they be given gunny sacks to collect bottles and cans for the rest of their lives and only allowed personal goods that will fit into a shopping cart with one wobbly wheel.

History of Water Vapor

Water vapor has an interesting history in the science of the atmosphere: In the 19th century physical science was in a compulsive quest for precision. Instruments delivered ever greater degrees of accuracy that all felt had to fit perfectly on the single line curves of mathematical equations. But nature doe not work that way: It wobbles. Scattergrams of data compiled to direct drawing single line curves were much more realistic than the curves carefully plotted by men in the unheated labs of the 19th century.

Engineers first came up with the idea of data "envelopes" with curves drawn around the dots to include all the data points feeling that more data would validate the idea the phenomenon existed within an area and not on one curve. Especially is this true in nature where nothing comes in a neat package until you get to the atomic level which is 100 million times smaller than the dimension we can perceive.

Water vapor is variable in the atmosphere and is the only gas that can leave air precipitating as a cloud, rain, sleet, snow or hail. Early atmospheric scientists ignored water vapor (1) because it was variable and (2) troublesome in the equipment, forming mists and causing corrosion.

As a result, all of the early work on air was done with "dry air," that from which most water vapor had been removed. Where water vapor is responsible for up to 99.9% of all atmospheric heating they rendered their work nonsense by design! All of what was written about the atmosphere was literally of no value for more than 100 years. Dry air is not the atmosphere, even in Death Valley or the Chilean desert, the two driest places on planet Earth.

Over 90% of Earth, the oceans and biome, 3% to 4% of the air is water vapor. We use three percent in our calculations as that is 100% humidity at 25 degrees Celsius and often the situation over most of America.

A molar volume of air, 22.4 liters, has a mass of 29.85 grams at Standard Conditions, zero degrees Celsius and one atmosphere of pressure containing up to 1.24 grams of water.

A molar volume is a molecular weight of gas in atomic weight units converted to grams. It has the same number of molecules for all substances. A "mole" of hydrogen has a mass of 2 grams where a mole of oxygen is 32 grams, but both have the volume of 22.4 liters and both the same number of molecules.

With 3% of air water vapor has 0.896 grams per 22.4 liters

and carbon dioxide with 0.038% is only 0.0113 grams. A mole of water vapor is 18 grams, but a mole of CO_2 is 44 grams. Thus we have 0.049 moles of water in a molar volume of air and only 0.000257 moles of CO_2 for 191 times as many water molecules as CO_2.

That there are hundreds of times more water molecules in the air than CO_2 molecules is important because the effects are on a per molecule basis, but anthropogenic global warming believers want you to think CO2 is the dominant substance in the system. It is not. This could only be true if molecules were under the intelligent control of tiny pilots or there were Divine or Satanic intervention.

Clausius Clapeyron and Antoine

If you see a lecture by a "global warming" or "climate change" promoter you may hear of Clausius Clapeyron. If the speaker is a Ph.D. he will display the equation on the board and where it is in Calculus few will understand it, but all will think he is "real smart." That is the point.

Water is the only substance in Earth's temperature range that exists as a solid, liquid or gas, and it can be all three in some situations. At the point where a glacier meets water is such a case. Clausius Clapeyron expressed it in a curve, but where solid, liquid and gas are three different states of matter three equations are needed which is where Calculus comes in as it is the branch of math that deals with such situations where algebra only deals with one at a time; solid, liquid or gas.

Calculus uses esoteric notation and conventions to define where certain of the components operate so the equations are long and impressive. However, impressive Clausius-Clapeyron does not work very well and was followed by the "Antoine" improvement equation, which is a variation, but not as elegant as "Clausius-Clapeyron" sounds, so all the guys doing presentations at Blue-Haired Bloody Mary and Gardening club luncheons put Clausius-Clapeyron on the chalkboard and the ladies are thrilled to be looking at a young, vigorous Ph.D. physicist talking about saving the planet in math they do not understand.

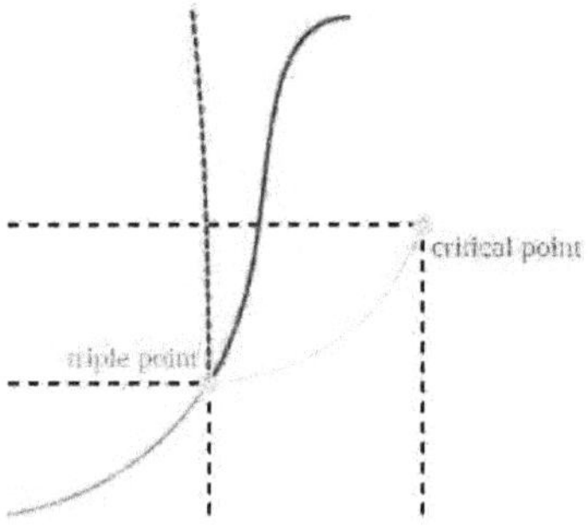

Clausius-Clapeyron Graph

Clausius Clapeyron predicts the pressure and temperature conditions where water exists as a solid, liquid or gas. The "triple point" is where water can be in any of three states depending on energy flow from one state to the other. The "critical point" is the liquid or gas, two phase point.

What does this have to do with "global warming?" Not one thing in the lower atmosphere where most of the heating occurs, but at the level where clouds form it is significant, but not the way the "man-caused global warmers" would have you believe.

They tell you that cloud formation, which happens often at 5,000 to 10,000 feet, eliminates the heating effect of water vapor and carbon dioxide takes over to control everything under it.

Not only is that not the case, but cloud formation works in opposition two ways: (1) Clouds are highly reflective and send solar radiation back to outer space and (2) the water droplets in clouds absorb higher energy waves in sunlight and re-emit as infrared for a greater heating gain than they would as water vapor. Of course this is never mentioned.

Furthermore, re-emission occurs in all directions so the water produced IR energy has a one chance in three of heading Earthward. Given random behavior by clouds, variations in coverage and depth, their full importance has been very difficult for climate modelers and they are responsible for more computer modeling errors than any other factor.

None of the many computer climate modeling systems have worked well enough to predict weather a few days in advance, but their authors have taken millions of Dollars in grant money claiming they could predict the climate for decades, even centuries! This is "fraud out of the box" and should be prosecuted for the criminal intent involved in every case.

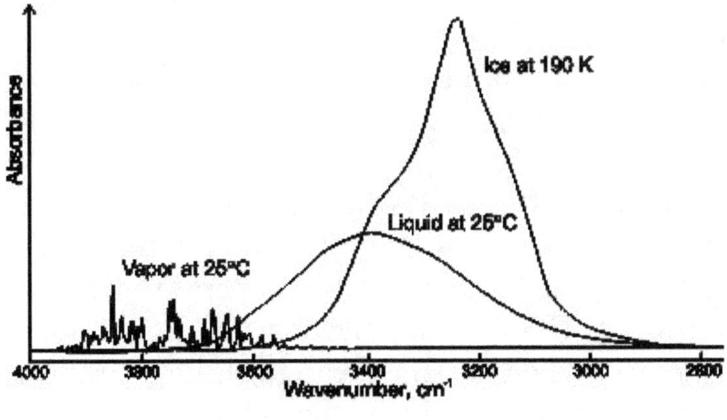

Cloud Energy Absorption

Smog and Air Pollution

Air pollution is not new. It has been a big problem since antiquity. Air in caves was smoky and cancer causing, but humans did not live long enough to get cancer in antiquity.

During the Middle Ages the diseases cholera and typhoid fever were common in Europe. Cities suffered with bad, smoky air. Epidemics arose from unsanitary conditions caused by human and animal wastes. In 1347, the bacteria Yersinia pestis, was carried by rats living on garbage and spread by rat fleas causing the "Black Death" outbreak of bubonic plague killing one third of the people.

Our greatest advances in public health have been made by purifying water, removing human waste and keeping one from the other. By the 1800s we understood unsanitary living conditions and water contamination contributed to disease epidemics.

Chicago built the first municipal sewage system in the United States to treat waste-water in the 1850's. While water purity was fairly well managed in late 19th century America, air pollution in U.S. urban areas increased well into the 20th century.

The Cuyahoga River in Cleveland, Ohio flows into Lake Erie and it became so polluted with oil and petroleum waste the water surface caught fire several times! The first fire occurred in 1936 when a blowtorch ignited floating debris and oils. The Cuyahoga caught fire several times over the next 30 years and was the subject of a hit Randy Newman song.

In 1969 a Cuyahoga river fire occurred that was very well covered by newspapers and magazines. Public response resulted in the Federal Water Pollution Control Act of 1972 often called the "Clean Water Act."

The Cuyahoga River fires provided motivation to create the Great Lakes Water Quality Agreement, establishing federal and state environmental protection agencies and created the Environmental Protection Agency, plus passed the Oil Pollution Act of 1990, which prohibits oil discharge in "navigable rivers," those large enough for commercial vessels and barges.

Air pollution from automobiles, industrial processes, and the burning of coal in factories and in homes had also been a serious problem. In Los Angeles the term "smog" was coined and much of it was caused by people burning their own garbage. The "LA basin" is a natural trap for smoke, but also a natural cauldron for the oxides of nitrogen made by ozone from sunlight's hard UV and cosmic rays. Add that to smoke from human activity and a really noxious atmosphere was the result.

The people of Los Angeles spent millions, if not billions, of Dollars on solving the problem and they succeeded by the 60's. Where automobile exhaust had been the big offender it has been rendered harmless by the catalytic converter that burned off the most offending combustion byproducts changing them to "harmless CO_2," as noted in the media.

In the 19th century, episodes of "smog," the combination of smoke and fog, in cities like London and Los Angeles caused many deaths. New York escaped only because it gets a sea breeze keeping the air clean.

Air pollution continued to be a significant problem up through the middle of the 20th century. In late October of 1948, 20 people were asphyxiated and more than 7,000 became seriously ill as the result of severe air pollution in Donora, Pennsylvania.

Like the 1969 Cuyahoga River fire, the 1948 Donora, Pennsylvania incident led to the creation of the Air Pollution Control Act of 1955. This was the first federal attempt to control air pollution. The Clean Air Act of 1990 set limits on the discharge of air pollutants from industrial facilities and motor vehicles, plus addressed acid rain and ozone depletion, two false issues from which we should have learned the nature of environmentalism.
These laws significantly reduced the quantity of pollution released into the environment. Grossly contaminated water and air have almost vanished. This has meant bureaus will have to find something else to regulate or reduce in size. That is one of the great drivers in anthropogenic, man-caused, global warming, now called "climate change."

The Ozone Hole

In 1972 UC Irvine Chemistry Professor Frank Sherwood Roland and graduate student Mario J. Molina were looking for a project with which they could get for a Federal grant. For reasons known only to them they picked on Freon™, one of the finest industrial chemicals ever invented.

Freon™ was widely used in refrigeration because it is not flammable and has excellent properties for such systems. Out-of-the-blue Roland and Molina accused Freon™ of "destroying the ozone layer" in a Federal grant application to the National Science Foundation, NSF, then awash with money for grants.

The proposal got immediate attention as it suggested new taxing and regulation. Very soon Roland and Molina were looking at a $60,000 check to get them started coming up with a plan, experiments and a paper. Who knew where this was going to end?

Ozone is formed throughout the atmosphere by ultraviolet light and cosmic rays striking oxygen molecules and split them to form two free atomic oxygen atoms, 2 [O]. It is popularly referred to as "O_3" but is actually single atoms of oxygen that are very short-lived and formed from the top to the bottom of the atmosphere. Then it forms brown oxides of nitrogen to make "smog." Ozone is formed down to a few thousand feet when the UV is all absorbed, saving us from ultraviolet wave burns. Ozone is longer lived in upper atmosphere levels where fewer molecules hit it, but there certainly is not enough to make a "shield." This is yet another "green" or environmentalist dream and myth.

Man can live no higher than two miles in altitude as 70% of all air is below that level. If we go up five miles, where commercial jets fly, 50% of the air is below us and we will die in a matter of minutes without air pressurization or oxygen supplementation.

At ten miles 90% of the atmosphere is below us and at 20 miles 99% is under us while at 30 miles altitude 99.9% of all air is below us. Space is said to begin at 60 miles. It is at the 30 mile altitude that the "ozone shield" people say Freon™ emits chlorine to destroy "shields" that protect us from ultraviolet light. But, with virtually nothing there, much less a shield, this is a lie that leaps from the pages!

The formation of ozone occurs from an altitude of 20 miles down to the surface and it does not form a "shield" where that requires molecules to be in contact, as they are in a solid or liquid. In a gas they are never touching other than in collision, a contact of 1/5,000,000th second we know as there is a "hiss" at five megahertz frequency you can hear on every short wave radio as the rate of these collisions are interpreted as radio waves of that frequency.

It is oxygen that protects us from solar ultraviolet and the process takes place throughout the atmosphere down to a few thousand feet.

Gases cannot form surfaces or "shields." That Roland and Molina got away with such a lie tells us much about "big science." The people at the top see money and power in the goofy idea gases form "shields" 20 miles up! In a sane, honest world they would have been laughed out of UC Irvine in an hour, but our world is neither sane nor honest.

Rowland and Molina had predicated their hypothesis on the work of Englishman Gordon Dobson, who invented instruments to measure ozone in the atmosphere through a virtual vertical column to determine how much ozone was in the field as it emits certain waves as it extinguishes.

Dobson hitched a ride on an IGY ship in 1957 and made note of the fact the ozone in the air was diminished in the Antarctic. He would have been alarmed if it were not as he was there in the winter when there was little to no sun so there should be little to no ozone in the air of his test column. A little ozone is formed by cosmic rays, which are always present. That Roland and Molina used this as a red flag also reveals they were panic-pushing or ignorant, but the National Science Foundation loved it.

DuPont, the inventors, patentors and producers of Freon™ were initially alarmed with the Roland and Molina ideas their product was destroying a mythical “ozone shield,” but then their lawyers reminded them the Freon™ patent was soon expiring.

They suggested patenting a new chemical and they could continue to control the refrigerant market if Freon™ were banned. Thus Roland and Molina were good news as far as Dupont was concerned and they went dumb while many scientists, engineers and technicians in the field went nuts with Dupont management’s absolute silence.

The people in the field knew only too well that Dupont had struggled to get Freon right and on their way had disas ters like the infamous Copa Cobana fire where 400 people died when the air conditioning went incendiary as the molecule Dupont sold was not totally halogenated as is Freon™.

Totally halogenating a molecule is more difficult than doing part of it, as was done with the early products

It is hydrogen on the molecule that burns and Freon™ has none. But, all Freon™ substitutes now contain the same hydrogen flaw as did all pre-Freon™ refrigerant gases. Every home refrigerator is now an incendiary bomb, but it was not when Freon was used. This is yet another instance where what government tells us what to believe is not true.

To justify their concept Roland and Molina wrote a paper postulating Freon™ in the upper atmosphere decayed to free chlorine and this was somehow blocking formation of the ozone in their mythical shield.

They were apparently not aware that nature generates 600 million tons of chlorine annually and this is 80,000 times as much as they claimed to be alarming! Volcanoes like Mount Erebus produce 1,000 tons of chlorine every day.

When Mount Pinatubo went off one million tons of the gas came out and were their concerns valid much eye damage would have been seen, but none occurred. Their paper was so poorly, or corruptly, prepared they ignored that natural sources of atmospheric chlorine were 80,000 times theirs.

They also failed to note ultraviolet radiation reaching the surface had been in decline for 50 years. Hundreds of scientists prepared a document, “The Tazieff Resolution,” stating the CFC/ozone issue was a fraud. Not one major news media organization gave it ink or air time showing our press had already been compromised politically.

During this time Dr. William Happer, of the Department of Energy dared to pursue this issue with a review of the data and proposed more accurate instrumentation. On hearing of his work Vice President Al Gore had him fired. As well, House Resolution 547 for an investigation of a kind that would have resulted in criminal prosecutions both died in committee and it was ignored by the press. This is more evidence that evil forces were in control of what we were told from 1970 on and probably much earlier.

The system of equations Roland and Molina proposed do not look probable in nature as they are endothermic, which means they need energy in places where it is unlikely to be. Nonetheless, Roland and Molina were given a Nobel prize for reactions that have never been observed in nature and no one has ever been able to demonstrate in any lab. Nonetheless, in “big science” these guys were big heroes because they got big money and monumental recognition.

The "R134a" molecule that replaced Freon™ is explosive, corrosive, poisonous and carcinogenic where Freon™ is none of those things. Requests to the EPA for toxic ity data on "R134a" are now ignored in spite of laws requiring it be given to requestors.

Detractors were asking why ozone depletion had not been seen in the northern hemisphere when most of the Freon™ manufactured had not been out of the northern hemi sphere. The windless "horse latitudes" at the equator stop anything released in the north from crossing. Same story in the south. The only Freon(tm) in the southern hemisphere was in refrigerators. Any effect should appear at the north pole.

NASA was asked about this and scheduled nine U2 flights over the north pole in search of chlorine ions in the upper atmosphere as predicted by Roland and Molina. The first seven flights found absolutely nothing. Then, the last two flights were directed to fly over Norway. That they found chlorine in that area was said to confirm that Roland and Molina were right. However…

The fjords of Norway are the only place macrocystis kelp is found in the Atlantic Ocean. It had probably been brought by Japanese mariners tra ding with Scandinavians when the Northwest Passage was open up to 6000 BC. It is common in the North Pacific, but not in the Atlantic.

This kelp does a unique trick of converting column VII ions to their elementary states. In other words it generates free chlorine and iodine, but retains the iodine and the chlorine bubbles off.

Every cell of that plant has one tiny crystal of iodine in it. We do not know why, but that NASA knew they would find chlorine in the air over Norway means they were digging very hard to find atmospheric chlorine in the northern hemisphere as this is a very obscure fact.

The Role of Water Vapor

We studied gas/energy absorption on the charts. When we saw the Tyndall absorption charts for air on the NASA website it appeared CO2 was a large absorber of infrared radiation. But, on examination we saw the chart had been inverted! Where the “0,0” intersection of the X and Y axes would normally be in the lower left corner, in the NASA chart it was in the upper right corner. It had been inverted and flipped. A casual reader or math-ignorant reader could be misled into believing CO2 was a much better absorber of infrared, “heat radiation” than it is.

We contacted NASA about this error, but our letters and emails were all ignored. We found another chart at the American Meteorological Society which was correctly drawn and we used it in our analysis.

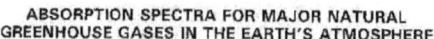

ABSORPTION SPECTRA FOR MAJOR NATURAL GREENHOUSE GASES IN THE EARTH'S ATMOSPHERE

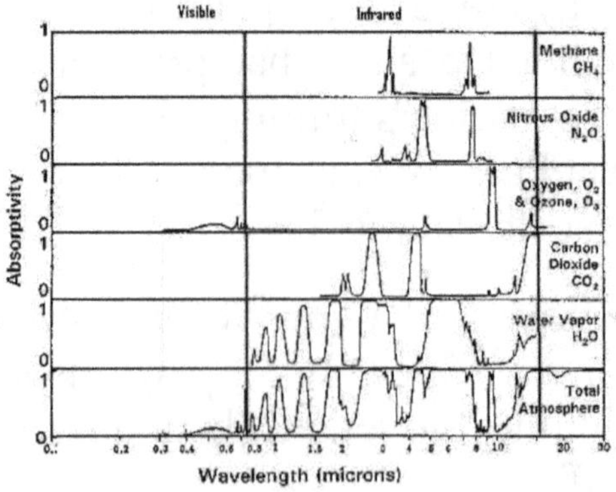

[After J. N. Howard, 1959: *Proc. I.R.E. 47*, 1459; and R. M. Goody and G. D. Robinson, 1951: *Quart. J. Roy. Meteorol. Soc. 77*, 153]

Absorption Area Analysis

The analysis was done by inscribing appropr iate geometric figures in the absorption areas; computing their areas, then summing them.

In a printout where this chart is 13 cm wide the gas areas between 0.5 and 15 microns have absorption fields with areas of 677 mm^2 as they are 61.5 mm by 11 mm.

Carbon dioxide has four absorption areas that are at 17.5, 28.5, 34 and 61 mm on the horizontal scale. Each is a triangular area we estimated in mm^2. Water vapor has six areas at: 2.0, 7.0, 24.0, 34.0, 51.0 and 63.0 mm.

The areas in the CO_2 absorption spectrum are: 7.5, 37.5, 22.5 and 37.5 mm^2 respectively while for water vapor they are: 19.5, 28, 64, 144, 240 and 35 mm^2.

The sum total of the areas covered for CO_2 is 105 mm^2, nine percent of the potential while for water vapor it is 467 mm^2, or 40% of it. On the basis of area alone the water vapor is 4.44 times the IR absorber as CO_2, but all waves are not created equal in terms of energy.

Energy Analysis of Absorption Spectra

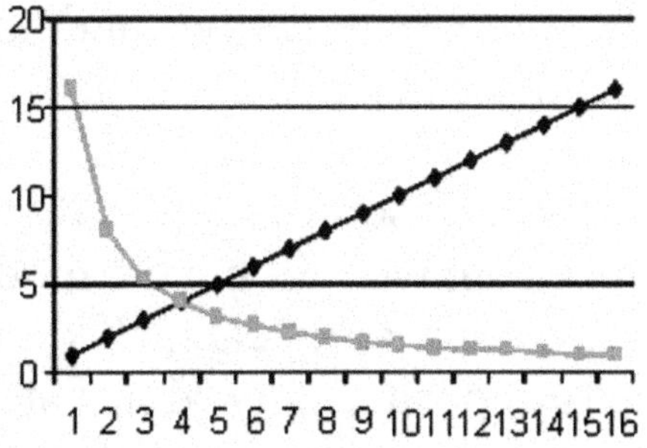

Wave energy is inversely proportional to wavelength and directly proportional to frequency. Energies are on the curved, pink line and wavelengths are on the blue, straight line. Quantities are on the vertical, "X" axis and the IR wavelength in microns are on the "Y" horizontal axis.

A one micron wave has twice as much energy as one at two microns. A wave of 0.5 micron length has 32 times the energy of one of 16 microns.

The mathematical equation, or model, for this is:

$$E = (h \times c)/w$$

E is energy, h is Planks Constant, c is the speed of light, which is a constant, and w is the wavelength. Here the numerator, (h x c), is a constant because one constant times another is a constant.

Analysis

As wave energy is inversely proportional to wave length we can compare these absorption areas quantitatively by arranging a scale where the wavelength is converse to the scale, i.e. the shorter the wavelength the greater the energy.

We here seek only to compare quantities and not label energy units as that complicates the concept.

If we analyze and sum the absorption areas dividing by their place on this spectrum we will have a model of the sun-atmosphere system correlating each absorption area with its energy-spectrum position in order that we may compare CO_2 to water vapor. We can do this with a set of equations noting each absorption area to its place on the IR spectrum in microns for CO_2 as:

$$[25/2] + [135/2.8] + [144/4] + [135/15] = 106$$

When we do the same for water vapor data we obtain:

$$[42.5/0.7] + [55/1.1] + [87.5/1.4] + [94.5/1.7] + [203/3] + [337/6] + [72/11] = 359$$

Thus the difference between water vapor over CO_2, IR absorption per molecule is:

$$359/106 = 3.39$$

Then, "How many molecules of each do we have?"

Carbon dioxide, CO_2, is now 390 parts/million, 0.00039 or 0.039%. A molar volume of air has a mass of 29.85 grams and thus contains:

29.85 X 0.00039 = 0.0116 grams of CO_2

And,

0.0116 g/ 44 g/mole = 0.000264 mole CO_2

Water vapor is 1% to 4%, depending on location, and we use 3% as a good working figure, thus:

29.85 x 0.03 = 0.895 g H_2O

And,

$0.895g/18g/mole = 0.04975$ mole H_2O

Thus,

$0.04975/0.000264 = 188$ X moles of H_2O/CO_2

This is important as absorption is also directly proportional to the number of molecules as the effect is a product of the relative absorption and the ratio of concentrations. Thus:

$$6.86 \times 188 = 1292$$

Water vapor is 1292 times more significant in "global warming" than carbon dioxide and is thus responsible for 99.9% of all atmospheric heating, but it is never mentioned in any article, textbook or film produced by publishers selling to public schools, in major media, by Albert Gore, Jr., Bill McKibbens or a federal grant supported scientist because water vapor is ubiquitous and cannot be controlled or taxed. It is just that simple.

The Keeling Curve

While Roger Revelle was the big gun in "global warming" a hot runner up is Charles David Keeling, also of Scripps Institute, and the hypotenuse of a two man, one woman triangle pushing man-caused global warming from Scripps. Did they know what they were really doing? No one can know, but if they did they deserve a special place in Hell for what they've done to America, science and education.

The total damage has now been estimated by Forbes to be $7.4 trillion Dollars to the economy and generations of scientists will work to overcome the doubt that has been created by those selling science for money when the truth is finally realized by the people.

The Federal government spent $320 billion promoting this fraud from 1988 to 2000 for 20,000 academic papers in search of a justification for this insane idea. They have sentenced two or three generations to paying double the tax rates anyone over age 40 paid through their lives or the economy crashes totally and millions may starve.

Many have long wondered about the Keeling Curve: It is so perfect: Experimental science is not like that. Mother Nature wobbles when she walks. But no one has dared challenge heroes of the "global warming gang." Some Federal officials have proposed prison or a death penalty for denying man-caused global warming!

Scripps was originally funded by private donations from widows of the ultra-wealthy San Diego retirees, but they found a deeper well of Federal IGY money to promote fear Earth could only be saved with ever-larger checks. Now they are part of the University of California, tapping public pockets for more money in return for panic pushing false science. The finest example is the famous Keeling Curve

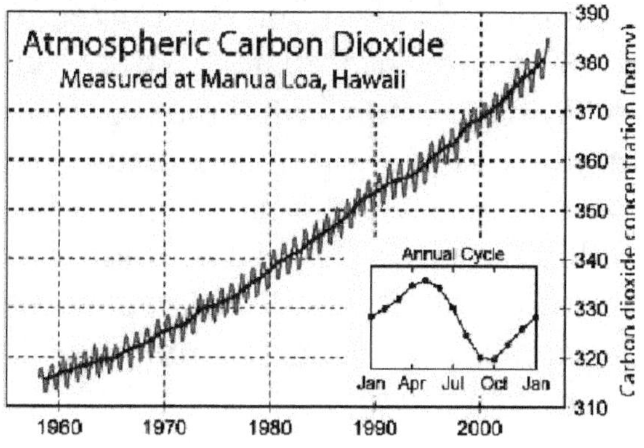

Charles David Keeling was a clever young man who loved the outdoors so much he turned down jobs, after getting a Ph.D. in physics, because they were in areas with climates he did not like. Physics has been a tough profession to get a job in so after not finding one in a place he liked Keeling became a post-doctoral student at Cal Tech and there he learned of the work of Dr. Roger Revelle reading his first paper on CO_2 concerns for the oceans.

Determining any gas quantities has long been difficult: There is so little CO_2 in air chemical analyses are both troublesome and unreliable.

In the chemical quantitative analysis of CO_2 trace amounts are captured in standardized sodium hydroxide solutions. A drop of dye is added, the color changes with pH, and the solution is compared with color standards to determine the change in pH and from that calculate the amount of CO_2.

The pH is "the negative logarithm of the hydrogen ion in moles per liter." Neutral is "7.0" as the hydrogen ion concentration is 0.000,000,1 or 10^-7 mole/liter, the usual situation for pure water.

Water contains a very small amount of ionized hydrogen as a consequence of molecular collisions. A few will dislodge one of the two hydrogen atoms on the oxygen of a water molecule, ionizing, or charging it, as a result. The actual number is about one in two billion, but in terms of moles per liter it is 0.000,000,1 or 1.0 x 10^-7 mole per liter. And in chemist's script it is "pH of 7," "the negative logarithm of the hydrogen ion concentration in moles per liter," where "-7" is the "negative log."

pH is the key to determining the amount of CO_2 in the

chemical quantitative analysis of air. A dilute solution of sodium hydroxide with a concentration of 0.01 mole/liter could be used with an air sample bubbled through it many times in 100 milliliters, 0.1 liter, of known concentration or the solution may be sprayed through the air sample at a low temperature repeatedly to capture all the CO_2 gas.

The devices to do this work are notoriously difficult to use and maintain. They have to be rebuilt continually as the caustic sodium hydroxide attacks the seals so they fail and the apparatus needs to be tested often to maintain validity.

In 1953 Charles David Keeling was a post doctoral student at Caltech, Pasadena, California, investigating carbonates in surface waters, limestone and atmospheric CO_2. This called for the construction of a precision instrument to measure CO_2 extracted from air and water without all the difficulties that were involved in chemical analyses.

By 1957 Applied Physics had developed an infrared gas analyzer that not only determined the gases present by absorption spectra, but their quantities. It would have to be calibrated with known samples confirmed by chemical analyses, but would reduce the complexity of an ongoing program considerably if their electronics worked well. However, Keeling wanted to develop his own syste m.

Keeling found significant variations in CO_2 concentration in Pasadena, and took his sampling equipment to Big Sur near Monterey. He took samples through day and night and on analysis in Pasadena noticed a diurnal pattern as plants produce CO_2 at night after making oxygen in the day.

Isotopic ratio mass spectrometry showed him the $^{13}C/^{12}C$ ratio in CO_2 at night was smaller than during the day as a function of plant respiration. This was to be expected as industrial activity and automobiles burn fossil fuel to make CO_2 that has no ^{13}C in it as it is made during exposure to cosmic rays in the atmosphere. Fossil fuel has not been in the atmosphere to be irradiated by cosmic rays.

More as a curiosity than not Keeling took samples for the ^{13}C isotopic test from Oregon to Arizona and saw the same pattern, which he should have expected if he understood plant physiology.

In 1956 David Keeling caught the attention of Harry Wexler of the US weather bureau and Roger Revelle at Scripps Institution of Oceanography where he proposed a global program based on infrared gas analysers to measure the atmospheric CO_2 concentration at several remote locations around the world including the South Pole station and at Mauna Loa in Hawaii. It all became part of the 1957-58 International Geophysical Year, IGY.

For the IGY project Keeling bought an Applied Physics infrared gas analyser, but then went on to develop his "manometer," like any man wanting to have a patent or patents as they can be very good for you many ways.

When he began his work in Hawaii he made a curious notebook entry in 1959. It read, "We were witnessing for the first time nature's withdrawing CO_2 from the air for plant growth during summer and returning it in each succeeding winter." Keeling was not a profound thinker. Meanwhile back in the laboratory:

Keeling devised an apparatus that would freeze CO_2 out of the air and the reduction in volume would equal that of the CO_2. “No wet chemistry and a direct measurement!” he likely heard in his physicist's mind.

Keeling used five liter air samples. At 280 ppm of CO_2 he would see a 1.56 milliliter reduction in volume in the tube after the CO_2 was frozen out of the sample.

The total CO_2 volume was 1.56 ml in 1958, rising to 1.90 ml by 2005. A difference of only 0.34 milliliter, or the volume of seven drops of water, over the 47 years but he claimed to detect year-to-year differences of 0.007 ml with perfect results! This is not possible as no apparatus was capable of that precision in that time. Nonetheless, Dr. Keeling managed to project a curve that worked for him!

Keeling's "Manometer Report III" describing the apparatus is published on the Scripps Institute website is available with the site search routine. Analysis of the mechanics do not confirm the CO_2 gas volume could be determined to 0.007 ml. The greatest such accuracy we have seen for systems of that kind in that time is 0.2 ml.

The Smoking Gun

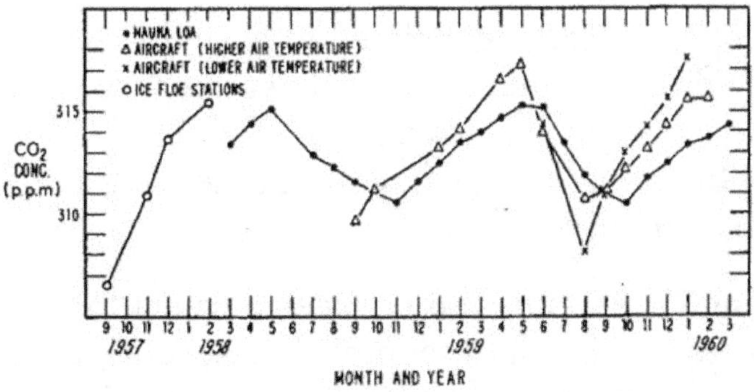

Fig. 1. Variation in concentration of atmospheric carbon dioxide in the Northern Hemisphere.

Tellus XII (1960), 2

The graph done by Keeling for the IGY project is a virtual "smoking gun" shows a seasonal change, but no rise and, in fact, a statistical decline from 1957 to 1960, but this does not appear in his famous "Keeling Curve."

While Keeling was said to be using the IR analyzer he makes no mention of it even in his later papers. It could be that he only wanted attention given to his invention, but why didn't he file a patent application? That is a standard practice for men in Keeling's position as patents publish new technology and are good for inventors professionally.

In the manometer Keeling made no provision for water vapor that would have frozen 78 Celsius degrees above CO_2 with a volume of 150 to 200 milliliters, and varying seasonally.

He never mentions dealing with water vapor in any part of the description of the process and it would be critical as most chemicals capturing water vapor also grab some CO_2 as there are molecular similarities needing to be accounted for in procedure and calculations.

For example: If Keeling claimed to have dried the sample with anhydrous calcium sulfate he would have introduced an error as this salt catches a small amount of CO_2. Same story for bubbling it through concentrated H_2SO_4, the most common alternative catcher.

It is telling Dr. Keeling did not seek a patent for his new “manometer” as he had accomplished several things never done before that would have been patentable if true. He was secretive about his work. A patent would protect it, but open it to critical review. He did not file for it and the most likely reason is that it did not work.

“Manometer Report III,” the last paper about his method, contains curious statements including: “The 1 cc chamber has never been used because equally good results have been obtained using the 4 cc chamber," and the mercury manometer meniscus reading "telescopes," as he call ed the scale magnifiers. They are not "telescopes." Please...

Keeling claimed, "readability of 0.002 mm and a precision of 0.005 mm," which beg the question, "Why are these two not the same? Then he later notes, "Each meniscus card consists of scale one mm in length with division of 0.05 mm set in a white below a black field."

This is confusing as the apparatus is shown as a set of vertical tubes, which would be normal, and he is calling for 20 graduations of a line one millimeter long. We have not seen anything with such fine graduations, which is not to say it did not exist, but without knowing the diameter of the tube holding the gas sample this note has no meaning.

Even more astounding: The Keeling procedure would have to take water vapor out of the air long before the CO_2 was frozen, but he makes no mention of it in a list of steps or procedures.

In Maui water vapor would vary between three and four percent of the sample and CO_2 0.028% to 0.038% over the term of Keeling's work, with water vapor up to 143 times the volume of CO_2 and an IR absorber Keeling makes no mention of! This is a breathtaking oversight.

If said "1 cc" chamber was never used then how did he know it would not produce better results? The "cc" term leaps from the page as that standard was changed to "ml" in the early 19th century when it was found the one liter standard block in Paris was flawed and had to be remade. Was he using the old standard? That is what he wrote!

It is curious that Dr. Keeling and Dr. Hansen prefer 18th century science terms to those of their own time. Keeling's reporting results one full magnitude beyond the capability of the apparatus is incredible and original sin in physical science.

Every physical scientist examining this work would see this, but no has one dared note it in print! And the most damning: While in Hawaii he did not make the case that all the new CO_2 in the atmosphere was coming from the burning of fossil fuel which would seem to be required for anthropogenic global warming alarmism.

In the year 2000 Dr. Keeling was one of six authors of a really strange, difficult to read, rambling, Calculus ridden paper entitled, "Exchanges of Atmospheric CO_2 and $^{13}CO_2$ With Terrestrial Biosphere and Oceans from 1978 to 2000." The paper is difficult reading even if you are science-trained as it rambles and has annoying language like "principal" where "principle," would be appropriate, "interannual" which is not a word, undefined units like "PgC yr^-1," and a strange "Conclusion" that reads:

"In summary, our principal finding from a new analysis of atmospheric CO_2 data is that interannual fluctuations in net exchange of atmospheric CO_2 are on the order of several PgC yr^-1 and correlate with strong El Nino events. The fluctuations clearly involve the terrestrial biosphere, and probably the oceans, but their amplitudes and phasings cannot be precisely determined. Calculations of these fluctuations, however, are not sensitive to further uncertainties which remain in calculating the long term sequestration of industrial CO_2 by the terrestrial biosphere and the oceans."

These 88 words take the prize for the obfuscation of any understanding and who knows how much we, the US taxpayers, paid for it. We have no idea what "PgC yr^-1" means as it is not defined in the paper. On this alone it would get an "F" in any upper division or grad course to say nothing of the other errors and confusions.

While he did do ^{13}C tests he did them strangely, never correlating them to man-made CO_2 in the manner that should be required for anyone making a case against the activities of man.

It has been claimed Keeling did $^{13}C/^{12}C$ tests to show all the new CO_2 in the atmosphere was man-made, a critical contention for the "global warming" alarmists, but we have not been able to find any beyond this paper that is fixed on El Nino while there is a more serious question regarding the role of the seas as the amount of CO_2 causing his readings is more than double produced by all mankind!

The atmosphere has a mass of 5.75 x 10^15 tons or 5,750 gigatons and of that 1.56 gigatons was CO_2 in 1957 when Keeling started his work and 2.18 gigatons the quantity when Keeling died in 2004. The atmosphere added 0.62 gigatons over 47 years; 13 billion tons per year.

Dr. George Woodwell did an analysis and inventory of the world's carbon for the January 1978 issue of Scientific American. He was really talking about CO_2, but for some reason chose to define it as carbon alone and grams as the unit of measure.

We can translate his figures to CO_2 with a conversion factor of (44/12) = 3.67 to relate it to our work. His real concern was that the figures used in discussing the CO_2 thought to have been in the atmosphere, and that known to be in the biome plus that thought to be in limestone do not "add up" by a very large quantity. A very large quantity of carbon is missing. We think it is in sea sediments as petroleum and if that is the case we will have a reserve that will last for thousands of years and incite thorough studies of the seas as sources.

Woodwell, and others, have concluded that all of man's activities produce about six gigatons of CO_2 annually while the natural decay of limestone in fresh water, dead plant and animal matter produce 160 gigatons of CO_2 annually. Man's contribution is only 3.22% and of that the United States makes 20% or 1.2 gigatons.

If we cut our production 80% per paranoid panic pushers we will have reduced the planetary output by 0.14%, an insignificant, undetectable amount and ruin our economy as a result.

We are the world's "breadbasket," exporting more grain than any other nation. Less efficient farming would return much more CO_2 thus the green gangs lose two ways! The curious thing about all this is that it is obvious in so many ways, but has gotten to the point where serious Democrats want a death penalty for deniers of "Climate Change."

The First Keeling Curve

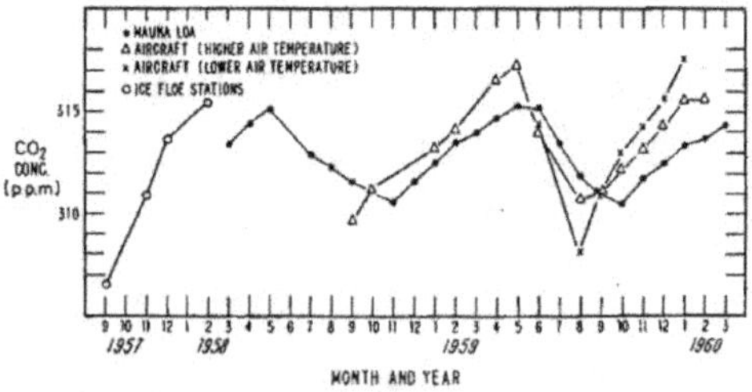

Fig. 1. Variation in concentration of atmospheric carbon dioxide in the Northern Hemisphere.

Tellus XII (1960), 2

The first Keeling Curve was part of an IGY report in 1960. It clearly showed no upward trend in atmospheric CO_2 for three years. It caused the end of Keeling's IGY funding. Soon after the version we are more familiar with appeared and was the Keeling stock-in-trade, his ticket to a long life of surfing and Dairy Queens. It is clear Dr. Keeling got the message with his first funding cut.

What Dr. Keeling documented was twice the CO_2 made by man so nature was "doubling down" for him. The most likely candidate is the sea as it is a massive sink of CO_2. The oceans have a volume of 1.33 10^15 liters of sea water, each of which can carry 1.6 grams of CO_2 per liter and a possible total of 2.13 x 10^15g of CO_2

Where Dr. Keeling was gathering evidence to indict the activities of man an upward trend would be a critical issue and it was his luck one was happening naturally and for reasons that have not been publicized.

The atmosphere contains 2.00 x 10^18 grams of CO_2, 1,000 times as much CO_2 so the global warmist's saying the seas are a great "sink" of CO_2 is just as silly as calling Earth a greenhouse. Said "sink" is a bucket compared to the atmosphere.

Liquid water is a very good absorber of the infrared energy from sunlight. Where less CO_2 is soluble in water at higher temperatures, any increase in the output of the sun will be accompanied by an increase in atmospheric CO_2, but again this is a bit of "dog wagging."

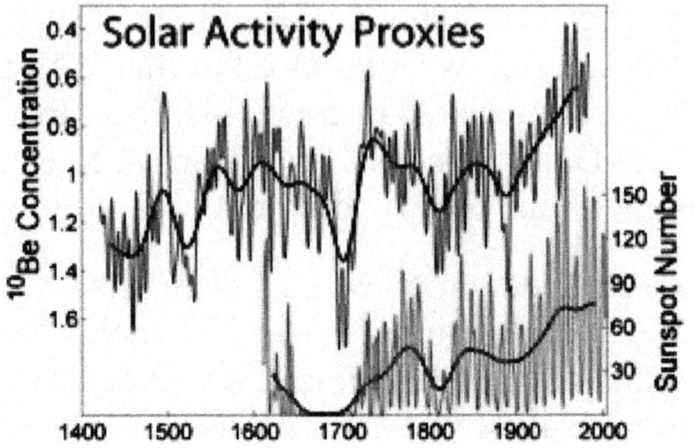

During Keeling's time the sun was on a rising output trend so it was driving CO_2 out of the seas. It can be quantified, but we do not wish to encumber this work with the full analysis. The point remains, CO_2 was being driven from the seas in the period Keeling was ostensibly tracking CO_2 with his "Manometer."

Geophysicists say about 1,000 volcanoes erupt on Earth every year and where 71% of the planet is covered by water it seems logical to expect 710 of these to erupt under water. They also produce CO_2, in addition to heating a lot of water, but just how much has not been determined.

The amount of CO_2 increase indicated from Keeling's, and other's work, is more than double that attributed to man from studies like those by George Woodwell. It would seem the highest likelihood is the volcanoes are putting out more CO_2 than we have attributed to them in the past.

Keeling's contribution is that he detected the trend and found a mathematical curve that worked perfectly; too perfectly. His "Manometer" was a great cover for his one equation that thanks to the sun, and continued activity of mankind, allowed him to make a perfect prediction.

Keeling was right in his overall, long term figures due to his good fortune to experience a warming sun cycle that put the increase on a curve that a linear regression would predict accurately. The "Exchanges of Atmosphere..." paper shows he was more interested in getting grants than doing honest research. And, thanks to federal funding he was a successful prototype for scientists who sell out for money. It all began with the 1957 IGY which shows us that science can go wrong with good intentions. There are evil men in white coats.

The International Geophysical Year

The International Geophysical Year was inspired by the International Polar Years, held in 1882–1883 and 1932–1933. It was to be held again in 2007–2009.

In March 1950, several leading scientists: Lloyd Berkner, Sydney Chapman, S. Fred Singer, and Harry Vestine, met in James Van Allen's home and decided the time was good to have a new worldwide "Geophysical Year" rather than wait for the Polar Year event considering all the advances that had been made in rocketry, radar, and computing.

Before year-end Berkner and Chapman proposed that the International Council of Scientific Unions should have an International Geophysical Year (IGY) for 1957–58, to coincide with an expected maximum solar activity cycle. The Eisenhower Administration announced the Vanguard satellite project as America's contribution.

Vanguard failed on the launch pad and Russia put up Sputnik with a military rocket while we argued over whether or not that would be "politically correct!" This was the first victory for that insane idea and it set off a national paranoia over science and our place in space.

Former Admiral G. Hyman Rickover wrote two books trashing American education comparing it to the private schools of Switzerland where rich people, who tend to have smart kids, polished their offspring for learning to rule the world, or at least own big pieces of it. This badly rattled American educators who in turn became political.

IGY triggered an 18-month "year" of Antarctic science. The International Council of Scientific Unions, broadened the proposals from polar studies to geophysical research. More than 70 existing national scientific organizations were involved.

Among them the Russian Vostok ice core project that drilled through the mile of ice covering Antarctica while collecting a 10 centimeter ice core that had layers as the ice was laid down seasonally and like the rings in a tree. The record they collected allowed them to slice samples from well defined times over a span of 440,000 years.

The slices included trapped air bubbles sufficient for gas analysis for CO_2 and the isotopes of oxygen which can be correlated with the ambient temperature at the time they were trapped.

Heavier gas molecules diffuse slower than lighter ones at rates that are inverse to the square of their densities by Graham's Law. Oxygen has two common isotopes O16, normal oxygen, and O18, heavier from neutrons added by cosmic ray collisions.

Where a molar volume of 22.4 liters for the usual O16 molecule will have a mass of 32 grams or 1.43 grams per liter one including on O16 and one O18 will have a mass of 34 grams/mole, and 1.51 grams/liter.

Per Graham's law the lighter gas diffuses at a higher rate, 1.51^2/1.43^2 = 1.11 and that difference can be translated into the temperature at the time the gas bubbles formed. To be sure the work is exceedingly delicate and subject to some question, but much testing is done to get figures in which we can have some confidence.

Scientists can estimate temperature from these differences. When this work was done on the Vostok ice core studies it revealed that over the last 440,000 years temperature has fallen and risen 21 times significantly. The amount of CO_2 in the air has risen 800 years after the temperature rise 19 times. The two times it did not were in the very early part of the record where data is less reliable: They are usually discounted as the record from there on is consistent.

Therefore, we say that change in temperature is cause and rise in CO_2 is effect. It is not the other way around and with over 400,000 years of data it would seem the idea of CO_2 caused "global warming" should have died a quick death, but it lives on for money and power; more than the elected ruling class has ever seen since the signing of the Magna Carta in 1215 AD, 35 years before invention of the Scientific Method by Roger Bacon and Albertus Magnus at the University of Paris, Sorbonne, in 1250 AD.

And so, "global warming," now "climate change" has a life of its own in the minds of socialists seeking new things to tax at ever greater rates.

Gore and Hansen

Albert Gore, Junior published "Earth in the Balance" in 1992 after having worked on it for 30 years and it is 400 pages of nonsense. He writes such profundities as "The internal combustion engine is a greater threat to mankind than nuclear weapons." That is classic creative stupidity and Gore gives no explanation for what he writes.

If we were to follow Gore's recommendations America would be able to support about 70 million people in a horse-drawn, sailing ship, candle-burning culture of the year 1900. A responsible man in Gore's position of being high born, in high elective office and able to get a book published, would hire a knowledgeable co-author and listen to him. Albert Gore, Junior is so bull-headed he would not do that, nor would he let an editor touch his work. As a result the book has errors in it that would have put principled publishing people off it totally.

To see page long paragraphs that are one sentence such that you have no idea what-the-Hell he is trying to say tell you this is a man who told the publisher not to touch a word, "It's perfect," and there are people in the publishing business stupid enough to abide by such idiocy coming from someone high born. We do have our own royalty.

Twenty years before "global warming" was postulated, in 1971, Jim Hansen, then of NASA, was asked to prepare a computer program that would predict a coming ice age as his chiefs thought one was coming as a consequence of all the smoke and particles kicked up in World War II, plus the volcanoes erupting annually. Curiously, they had the idea that carbon particles would cause some global cooling and should be the heart of the model.

Computer climate models are essentially big spreadsheet programs with routines written to make the air mass of one cell move to the next per unit time. If there is a barrier, you just omit the link to the next cell. The prevailing wind patterns, with known velocities make this a fairly simple design job, but it gets complicated fast as factors like clouds over water or land are added. And, with the daily turning and tilt of Earth it can get messy as time is part of the picture.

Generally, one-third of the sky is overcast with clouds that normally cut the illumination, and energy up to 50%. This is not the only variable and the model is never right. We may wonder why anyone attempts to write them, other than that they get paid a lot of money for them. Taking these systems seriously is a joke.

Jim wrote the program, in all likelihood on a spreadsheet system like Excel and got it to predict an ice age so the group made a presentation to a committee of Congress. The elected folk were utterly unimpressed as they did not see any new taxing or bureau building opportunities in having a 5000 foot tall glacier slide over Canada and bulldoze us as one did 10,000 years ago. They did not like the story and told Jim Hansen and his friends exactly that.

Jim went back to his desk to think for 17 years and as if he were some kind of locust, which he resembles, heard of Roger Revelle's "atmospheric heating" concerns and Charles David Keeling's never-ending Maui vacation with Dairy Queen cones paid for by rich widows of the Hotel Coronado Blue-Haired Bloody Mary Clatch. "Bingo!"

It is unknown how Albert Gore, Junior and James Hansen got together, but we are confident the first meeting was in a dark corner of a DC parking garage. And, that Jim had jiggered his old "Ice Age" computer program to produce a global warming outcome with a little numeric tinkering.

There are two kinds of geniuses: The kind you read about that have great flashes of insight suddenly and the kind that have very slow awakenings. Jim Hansen is the latter kind and Albert Gore, Junior is not in the building.

It took the better part of a year for Jim to explain to Al what he had in mind and how to pull it off. Al is not your-better-than-average-quick-study-politician. Al is “thick,” but Jim stuck with him and "the rest is history!"

On the hottest 1988 day in Washington, DC history Albert Gore, Junior and Dr. James Hansen made a presentation to a joint Committee of Congress on Science, populated by elected folks who knew nothing of science. The content was on CO_2 from fossil fuels heating the atmosphere and how they should be regulated and taxed! Bingo! They said the magic words! Nothing has been the same since.

Dr. Hansen was so thorough he had Albert Gore, Junior arrive early so they could open all the windows and shut off the air conditioner, later claiming it was a malfunction of the machine. Liberals are good at the blame game.

The committee members were stunned, and drowning in sweat, but came away with Champagne dreams and Cavier taxes dancing in their evil heads. It was a great success and before it was over Dr. Hansen had his own little bureau on the campus of Columbia College in New York City, home of US Socialism, a speakers bureau agent and millions of Dollars in honoraria forthcoming.

When people ask about water vapor James Hansen talks about "forcing." It is a resurrection of Vitalism, a Divine Theory of Chemistry, because it stipulates water may not evaporate until CO_2 gives permission or drives it into the air. It is utter nonsense, but no one challenges Jim. He is careful never to write of it in his papers for fear that a peer reviewer would declare him stupid, insane or a scoundrel.

Water evaporates very well by itself. Inputting the string "global warming water vapor" to Google will yield over 100,000 websites explaining the phenomenon. However, all are ignored by government scientists and politicians. This is done because taxing and regulating carbon is the greatest opportunity the elected class has had for more power and tax revenue since the Magna Carta of 1215 AD that ended the Divine Right of Kings and now the science community sees it as a chance for more grant funding. “It’s all funding!” they scream on getting Ph.D.s.

Adding carbon dioxide forces water out of the atmosphere because CO_2 is a true gas at Earth temperatures, boiling at -78 Celsius degrees. Water is a vapor that precipitates as it boils at 100 Celsius degrees, 178 degrees up the scale from CO_2 and vaporizes in the range of temperatures on Earth.

If air is saturated water vapor precipitates to liquid if the pressure or temperature fall. If it acquires smoke or dust particles on which it may accrete or a shock wave from a bullet, explosion, aircraft propeller or rocket pass through air creating a fast high, then low pressure wave the vapor precipitates to be seen as mist, a white trail or cloud.

The real nature of the atmosphere is embarrassing to the anthropogenic global-warming gang. Cooling has been caused by light filtering of particulate matter from fires, smokestacks, battlefields, bombed locations, volcanoes and meteors, but an increasing concentration of carbon dioxide producing a lower concentration of water vapor, which is a far better IR absorber is a fact. The idea of increased CO_2 causing air heating on Earth is not only unsupportable but laughable in view of Le Chatelier.

It has been postulated by Jim Hansen, and others, that CO_2 produces a layer in the upper atmosphere that reflects IR, infrared, radiation back to earth. This is crazy two ways:

Reflection requires a surface. Gases do not form surfaces. Only liquids and solids form surfaces that can reflect as their molecules and atoms are in contact, as required.

CO_2 weighs 44 grams per mole. All gases have the molar volume of 22.4 liters. At the surface, that volume of air weighs 29.85 grams as it is a mix of nitrogen, oxygen and water vapor plus the trace gases, one of which is CO_2. It is 47% heavier than air, diffuses slowly and remains at lower levels where it needs 2.45 times as much energy to diffuse per Graham's Law that gas diffusion varies inversely with the ratio of the squares of the gas density.

The per liter density of CO_2 is 44g/22.4 liter or 1.96 while that for H_2O is 18g/22.4 liter or 0.80 and the square of 1.96 is 3.84 while the square of 0.80 is 0.64 and 1.96/0.80 equals 2.45 so CO_2 molecules require 2.45 times as much energy to rise as water molecules so we should expect to see 1/2.45 = 0.408, about 41% less of them than water vapors at the higher altitudes. Therefore, the case for CO_2 being a major player in the atmosphere weakens the higher you go.

The atmosphere cools at higher elevations and by 5,000 feet most of the water vapor has condensed to form clouds and often they are of ice crystals. This actually increases the heating effect of water in air as ice and water capture the blue through violet and beyond quanta. They then re-emit these as infrared, IR, heat waves so again water plays a more significant role than CO_2 in the atmosphere.

It has been postulated by Jim Hansen high altitude CO_2 captured IR energy is re-emitted Earthward, but not to outer space! This claim of a kind of molecular intelligence or direction orientation is insane. Molecules re-emit in all directions randomly and we can analyze this in geometric terms to see just how nutty a professor is Dr. Hansen. We locate objects in space in three dimensions on the three planes of the "Cartesian axes," x, y and z. To the classic "x,y" axes of Algebra we add a "z" axis, third dimension as molecules re-emit in all directions in three planes. They do not direct their emissions to Earth by tiny pilots, Divine or Satanic intervention as would be required by the Hansen hypothesis.

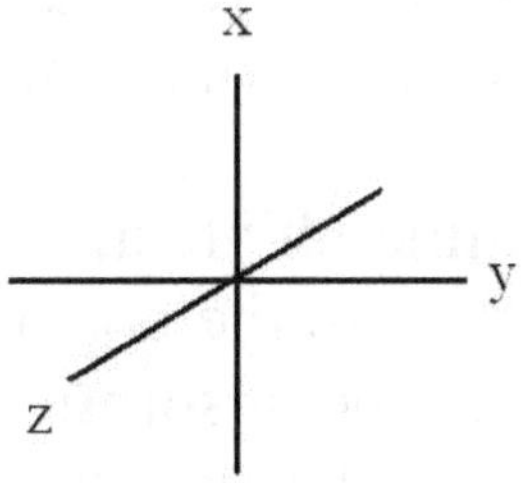

We divide circles inscribed on each plane in degrees for simplicity as we cannot illustrate or perform operations with infinity which is the real number of directions on a circle, but such operations are undefined in mathematics.

For convenience, and convention, we have (360)^3 or 46,656,000 directions in three dimensional space for any molecule in terms of degrees. If we assume a molecule is going to emit an IR quanta it could be re-emitted in any of

46,656,000 directions, in terms of degrees.

If we assume the molecule is at an altitude of 20 miles it will have an earthward window no more than 120 degrees in each of three planes 120^3 or 1,728,000 paths, in terms of degrees and the probability of any one path is:

1,728,000/46,656,000 = 0.037 or 3.7%

This is one chance in 27 and not supportive of yet another Hansen hypothesis. We have to say the academic Senate of the University of Iowa should rescind his Ph.D. "For crimes against science and education." This man has done monumental, perhaps irreparable, damage to America.

Another concomitant insanity was the claim that heat was escaping at night. These "scientists" have to be reminded the sun is not shining at night so there is no driver. Night emitting molecules in plants are in complicated chemical systems unlike the simple ones in the atmosphere.

They had previously claimed high energy wave capture and re-admission of IR was in the upper atmosphere, but experiments show 99% of the absorption and heating is happening in the bottom of the atmosphere making most of these ravings little more than sanity hearing testimony.

We're All Gonna Drown!

In May of 2014 the respected science journal "Nature," the "Washington Post" and the "Daily Beast" all reported that an Antarctic glacier melting would lift sea levels four feet. That did not seem right and a few minutes with the back of an envelope and pen we confirmed this was nonsense.

Earth has an 8,000 mile diameter thus a surface area of 201 million square miles by A = 4 pi r^2. Earth is 71% covered by oceans with average depth of 12,232 feet, 2.31 miles, thus 143 million mi^2 times 2.31 for 330 million cubic miles of seawater.

To determine the effect of melting glaciers we only need to know the average depth of the oceans and with the depth we can determine the amount of seawater per foot. Adding that amount of water raises the seas one foot.

The authorities and geography literature say the seas have an average depth of 12,232 feet. Thus, for every foot of depth there are 62,536 cubic miles of sea water and it will take that much water to raise the seas one foot.

According to NASA the glacier they claim will raise the sea level four feet has an area of 255 square miles and it is one-third mile deep. This gives 84 cubic miles of ice, but ice loses 10% of its volume on melting so it is 75.6 cubic miles of water. Therefore, the change in depth would be:

75.6/62,536 = 0.00121 ft or 0.0145 inch

This is about the thickness of a human hair. But all the "greens," and panic pushers, claim we are going to melt all the ice on Earth. Geophysicists say they cannot imagine a circumstance where all the ice on Antarctica will ever melt as it is in the dark six months of the year and unlike the Arctic, the ice is on a continent, not in a sea thus cannot gain heat from ocean currents. It is high, dry and cold.

Ice melting in the Arctic ocean has no effect on sea level as it is floating and like an ice cube in a glass melts with no effect on the water level. It is already displacing water to the same volume it will have when melted. It is floating because ice is 10% lighter than water. The only ice we can melt and contribute to raising the seas is that on Greenland and in the glaciers.

The America Geophysical Society says the increase will be about 17 inches which is in good agreement with my w ork.

The US Geological Survey claims the seas would rise 217 feet if all ice melted but they include both poles and where the South Pole is in darkness six months of the year AGS geophysicists say that is not possible.

The Third UN IPCC report claimed sea levels would rise 24 feet, but don't publish their methodology. We suspect there was none. This appears to be more of their "arm chair science" and the embarrassment does not end there.

This is a problem that would take a well taught junior high school science club 15 or 20 minutes. Today it is doubtful our college science majors could do it as science education has become environmental indoctrination. It is the greatest tragedy of our time. Science sold out for money.

CO_2 and the Sea

Jane Lubchenko

The spirit of a young Dr. Roger Revelle now occupies Jane Lubchenko, who is the Head of the National Oceanic and Atmospheric Administration (NOAA). She has said the sea is "…climate change's equally evil twin" for the consequences it may have for everything from the navigational systems of spawning salmon and the health of coral reefs.

Dr. Revelle invented "anthropogenic global warming" while teaching at Harvard and gave Al Gore a "D" in the only science course he ever took. He has been seeking revenge ever since.

Revelle had been the Director of Scripps Institute of Oceanography and author of a paper speculating effects on the seas from increased CO_2 in the 1950's. He found there was no effect, but writing that it should be a concern got him the Directorship of the American Association of Scientists who could smell money in Revelle's allusion.

Revelle's paper documented the seas 330 million cubic mile volume have a greater affinity for CO_2, by 10%, than fresh water and a pH of 8.0 to 8.2 making it very slightly basic, hence the greater solubility of CO_2 due to forming bicarbonate ions, HCO_3^-. Gases that react chemically with a solvent are much more soluble in that solution as it involves molecular combination.

This happens only slightly in fresh water. A pH of 8 means there are less free hydrogen ions in the sea by 80 billionths of a mole per liter, which for hydrogen is 1 gram, the mass of 20 drops of water. The 8.2 means 82 billionths gram. The difference much less than microscopic. It is atomic! Nonetheless…

In a UCTV appearance in the early 2000's then Director of

Scripps, Dr. Tony Haymet became very emotional over this change. I emailed Dr. Haymet my analysis and he did respond with a long argument "...even though very small, it is still significant to sea creatures." After seven years doing underwater photography all around the western hemisphere, doing much technical reading on the sea, with the perceptions of a Chemistry major, Biology minor, I say not true.

Every liter of sea water contains 1.65 grams of CO_2 and there are 18 X 10^20 liters in the oceans so there are 29.7 x 10^20 grams of CO_2 in the seas. This is 20 million times the annual output of CO_2 from all sources on the planet of which man is only 3.6%. Concern for man's CO_2 in the air or oceans is utterly insane!

These facts are increasingly apparent to the science community. It is only a matter of time before conscience becomes the driving force in the ghost of the seas case. Even before that, we have hope science will come to its' senses as the education of scientists has become politicized in the United States and the quality is in decline.

CO_2 and pH

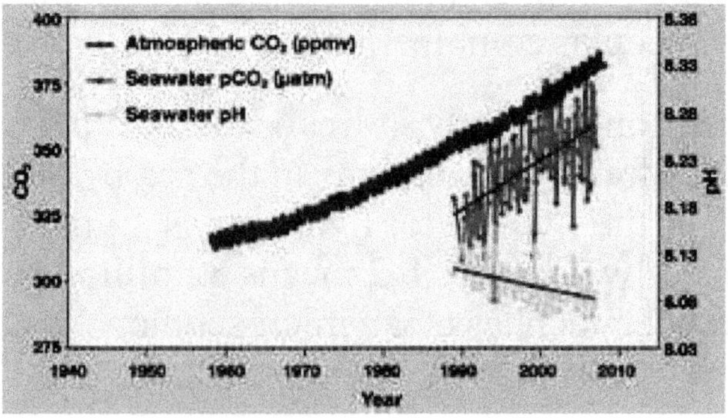

Feely-Sabine Graph

In a 2004 paper two NOAA Pacific Marine Environmental Laboratory senior oceanographers, Dr Richard Feely and Dr Christopher Sabine, declared increased atmospheric CO_2 causes a reduction in seawater pH. In the paper they wrote, “The impacts of ocean acidification on shelled organisms and other animals could negatively affect marine food webs, and, when combined with other climatic changes, could substantially alter the number, variety, and health of ocean wildlife. As humans continue to send more and more carbon dioxide into the oceans, the impacts on marine ecosystems will be direct and profound.” And, “The message is clear: excessive carbon dioxide poses a threat to the health of our oceans.”

Meanwhile, Mike Wallace, a 30 year career hydrologist and PhD candidate at the University of New Mexico reviewed the paper showing a strong correlation between rising atmospheric CO_2 levels and falling oceanic pH levels noting the chart began in 1988. But, Wallace knew oceanic pH measurements dated back 100 years and that data had been ignored for computer model projections. This was suspicious and he found data to refute it.

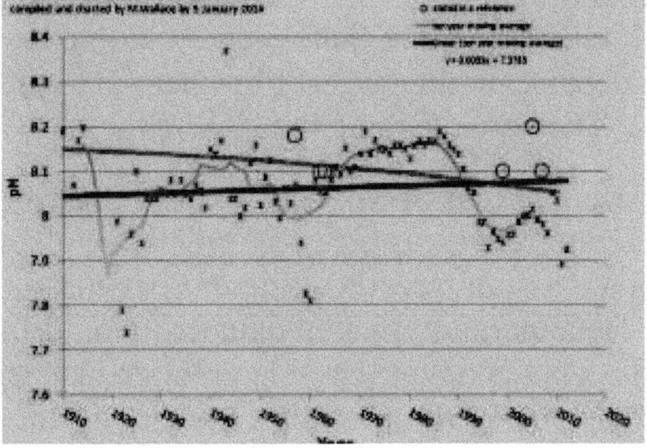

The Wallace Analysis

Wallace emailed questions to Feely and Sabine and they said it was inappropriate for him to impugn the "motives or quality of our science" warning "You will not last long in your career." They said the data Wallace found was not relevant to the issue.

Wallace also got original instrumental records from which Feely and Sabine had cut their graph of doom and plotted a time series chart of his own, covering the period from 1910 to the present. This showed clearly global acidification was an invention of Feely and Sabine. There had not been such a reduction in oceanic pH levels in the last century.

Oceanic pH is not an issue as long as it stays higher than seven and given the amount of CO_2 known to be in the seas it is impossible that anything mankind could add would ever be noticeable as we make but 3.6% of all the CO_2 produced annually on Earth. By far most comes from the decay of limestone and no freaky federal functionary can do anything about it!

Between calcium rocks known to be in the seas and consumption by plankton, seaweeds and kelp, oceanic production will little more than keep up with consumption. The oceans are the largest bio-chemical system on the planet. They are a large CO_2 sink.

This is a case like that of Michael Mann's drawing his cataclysmic inferences from a very small sample of tree rings from a Russian forest he did not visit and sample properly. He apparently picked data that would fit his hypothesis. That is not acceptable practice in scientific research and he has been roundly criticized for it.

It will be interesting, and telling, to see whether or not Mr. Wallace gets his degree as so much of academia has sold out for Federal money which ironically was taken from the people it is now ruining. Again, government has brought us a case of environmental fraud.

97% of All Scientists

Having heard "97% of all scientists believe in man-caused global warming" we went in search of the phrase and were not surprised to see a reference to Dr. Naomi Oreskes, Librarian, Scripps Institute, whose 2004 study concluded that "There is a consensus among scientists anthropogenic global warming is real."

Scripps Institute was the birthplace of global warming thanks to the works of Roger Revelle and later Charles David Keeling. It is an amazing coincidence to which Dr. Naomi Oreskes, PhD Science History, adds her name and fodder for occult hypotheses of evil points on the globe.

Naomi claimed her survey of the 928 peer-reviewed papers on "global climate change" journal articles between 1993 and 2003 was the ISI Web of Sciences database and that it documented her conclusion.

A dubious Dr. Ben Peiser, of Liverpool, John Moores University, UK, examined her work, and found it had been done by students. This is forbidden by all professional journals, and there were 12,000 papers in the full ISI database not 928! Dr. Oreskes students were told to read only the abstracts supporting anthropogenic global warming. This alone disqualifies the study completely.

We have to note that Dr. Peiser came under such pressure for his having rattled the windows of the hallowed halls of so much governmental funding he recanted his original contentions, but the facts he reported cannot be changed. The Oreskes study was fatally flawed.

For a rare time in their 100 year history the Editors of SCIENCE magazine published an erratum stating the standards Oreskes applied were not valid! This should have been the end of Naomi Oreskes in academia, but she was right at home at Scripps Institute, the home of Revelle and Keeling. Then, Harvard College offered her a higher position and she accepted!

Examination of the Oreskes work by Lord Monckton who actually read all the abstracts, revealed less than half of the 928 articles agreed with the Oreskes' "consensus" claim to say nothing of her claimed "97%!"

Before the SCIENCE errata, which did not appear for six months, a University of Chicago Masters degree student, Margaret Zimmerman read the Oreskes paper and got the bright idea to survey the 10,257 members of the American Geophysical Society, earth scientists on the question.

Margaret's thesis project advisor, Dr. William Doran was in favor of Margaret's idea while neither of these nimble noggins realized such a survey would cost over $10,000 in printing and mailing alone, plus a professional team of tabulators and analysts to be done in a reasonable time.

At some point they reality struck: Dr. Doran put his name on the report as Margaret wanted to seek publication and he was interested in a free ride, even if the work was not valid.

If you read the final report carefully you will find that they claim to have "thrown out" 10,180 of the responses which left them 77 of which 75 were positive for a 97% score! "Well, lookee here! Agreement with Naomi! How about that?

The truth of the matter is likely they did finally figure out what this survey would cost so Margaret picked out 100 project Directors and Chairs with hiring responsibility hoping for a job offer. 77 responded and 75 agreed, unless they simply made it all up, but let's give them benefit of the doubt for 100 mailed questionnaires, the printing, mailing postage and return postage would have cost about $300.

Directors and Chairs are in their posts because they are politically and financially savvy. We will bet the farm these guys were all looking for Federal grant funds and believe Uncle Sam is Santa Claus. "Global warming" is a dream come true for such men.

Ms. Zimmerman wrote of her study: "In our survey, the most specialized and knowledgeable respondents (with regard to climate change) are those who listed climate science as their area of expertise and who also have published more than 50% of the recent peer-reviewed papers on the subject of climate change, 79 in total. Of these 96.2% (76 of 79) said temperatures had risen since the 1800's and 97.4% (75 of 77) answered "yes" to "human activity" as the cause. Thus she quoted "97%, the mean of 96.2 and 97.4.

This is the basis for the figure that rolls off the lips of President Obama, Al Gore and other anthropogenic global warming promoters. The original survey and notes were available for a few months as a free PDF file at: http://tigger.ulc.edu/~pdoran/012009_Doran_final.pdf., but is has been taken down. We can only wonder why?

This "study" was widely reported by CNN and "Frontline" who referred to the 10,257 person sample and not the 77 participants actually counted when in fact it was clear 99.3% of those queried did not agree man causes climate change if we believe they spent more than $10,000 on the project. And, that is simply not possible.

Several science writers were curious about this point and when asked Margaret and William said they had done it by email. There is no record of any follow-up or evidence of any kind offered so we doubt the story, but even if true it speaks to lie of their conclusion.

If an academic receives a query on a controversial subject he will ignore it as a matter of course. That they claim to have thrown out 10,180 of the responses and focused only on the real "climate scientists" why was that decision not made before they spent what would have been several to more than $10,000 whichever way they did it? Their story makes no sense any way it is examined.

Liberals fall in love with certain numbers: During the Clinton era it was always "400 economists agree," or "400 historians confirm," or "400 nuns on bikes think." In this is case it is "97% of all scientists" when 97% of scientists have never agreed on anything, but getting more money in ever larger grants. Liberals have poisoned science with their 97% fraud with the justification "It is more money for science." We say it is money not only wasted, but that which has poisoned science education now such that many science industry employers have found they have to hire foreign trained scientists and engineers as ours cannot do the problems!

The Economics of Global Warming

President Obama has said we would be much better off economically if we eliminate "fossil fuel" claiming it is responsible for all kinds of evils, costs and problems. As it is true coal firing plants output chemical waste, much of it can be captured with Cottrel precipitators, 100 year-old technology, but capturing metals like mercury, uranium, trans-uranium and rare earth elements is difficult as there is very little of them in coal ash. Mercury is the only real threat to humans.

In a critical analysis much of the concern over many of the elements is overblown as they are in very small quantities that are widely dispersed and of no threat to people, plants or animals we use for food. We have a far greater threat in fish that are caught in lakes and rivers that are in or flow through soil having these common elements.

President Obama says "renewable energy will boost the economy." Hydroelectric is the cheapest such power, but there are only so many places we can build dams and the "Greens" get all upset when we do in spite of the fact one of their favorite little critters, the beaver, is a dedicated dam builder and has saved many areas in the west with their amazingly well made dams, ponds and lakes.

There are several sources of electricity including coal and natural gas power plants, nuclear and geothermal. It would seem geothermal will become more important as we can have it anywhere we can drill a 10,000 ft. deep hole.

Let us examine the Presidential proposition and determine whether he is correct or just blowing us more smoke. One chart tells us the whole story:

Energy Costs Comparison	
Resource Type	**Average Cost (cents per kWh)**
Hydroelectric	2-5
Nuclear	3-4
Coal	4-5
Natural gas	4-5
Wind	4-10
Geothermal	5-8
Biomass	8-12
Hydrogen fuel cell	10-15
Solar	15-32
Sources: American Wind Energy Association, Wind Blog, Stanford School of Earth Sciences	

After hydroelectric is nuclear, which is amazing as the costs of regulation and approval are utterly phenomenal. Immediately after World War II we had an opportunity to build many small, self-contained and automated nuclear power plants that could serve small cities or parts of big ones with high efficiency and safety as they would not contain enough fissionable material to make a weapon.

Two forces normally opposed to one another convened to kill this movement: Big Buck capitalists and "Greens" in a shotgun wedding of convenience. The Big Buck gang was out to get as many "rate payers" as they could per power plant and to the "Greens" this was just a step for them in killing nuclear power as they planned for it to collapse with regulatory pressure and economics of scale. Both of these evil groups won and the people lost as transmission losses waste 50% of all power.

Now we have a gigantic power grid that is a single target for terrorists instead of many small systems they could not affect greatly in a single attack. Again, the elected ruling class has failed us completely.

Coal and natural gas are next on the list of economical energy sources coming in with the same costs while the difference is the market price of fuel. Price is a product of cost, availability and demand. Given all factors coal and gas are our most stable energy sources. That in itself is a big reason to invest in them: They will always be there when we need them.

Wind power has many flaws: The first is that the wind does not blow all the time. We need electricity at the times it is most unlikely to be blowing; at night. Plus, wind and people do not get along. People do not choose to live where the wind blows as a result the "wind farms" are always tens of miles from the places they serve and the line losses are 50%, doubling the cost of the power. As well, wind power users need to have storage capacity. This reduces efficiency as alternating current can be transmitted at low cost, but it cannot be stored. It must be converted to direct current that can be stored in electrochemical cells called "batteries," that are very expensive and have limited life. They need to be replaced every three to five years.

Wind mills have been criticized for killing many birds, but it may be they were only killing the birds stupid enough to fly into them and that windmills will increase the intellect of birds sufficiently to where they can apply for Federal benefits on one of our many programs. Perhaps they can be trained to vote. Is this why Democrats really want more windmills?

Geothermal is only happening in North America at one location, The Geysers in northern California about 100 miles north of San Francisco straddling Mendocino and Lake counties. This field outputs enough electricity to fill the needs of the San Francisco Bay Area, but it is widely distributed and 50% is lost warming bird feet in the grid.

With fracking technology it may be possible to drill steam wells just about anywhere as the rocks at 10,000 feet are hot enough to generate steam. With fracking to create a cracked structure into which water could be injected will turn those areas into steam producers. The steam will be at sufficiently high pressures to drive electric dynamos and the steam can be condensed into pure distilled water.

This plan could result in communities that would use very little water as all the waste water they produce would be pumped into the deep, hot rocks, vaporized, condensed and recovered. They would need only enough water to replace that lost to evaporation and that could be virtually nothing.

Biomass means the production of methane gas by the decomposition of digestible organic material. It works, but is not economical given everything that has to be done to and with the gas before it can be used as fuel.

Hydrogen is an even worse fuel given hazards that come with it. Where it is so light it has to be under very high pressure. This means all the containers, connections and piping have to be made to high standards with high quality materials and be handled by highly trained personnel.

The greatest danger of all is fire as hydrogen flames are invisible in sunlight and nearly impossible to extinguish. So we have in hydrogen a gas that is more expensive than any other fuel gas, produces less energy per pound and requires more expensive, technically demanding systems making it uncompetitive to any other gas.

The big loser in the energy competition is solar and for very simple reasons. The panels have to be covered with glass to protect the delicate solar cells. These protective glass sheets cannot contain any iron as that will block the higher energy short wavelengths in sunlight. That can be prevented by inspecting the glass on end from where they should look blue or clear for sunlight passing through. Iron causes the glass to have a green cast on the side.

The real problem is simple physics as any polished surface is a 100% reflector at any incident angle below 45 degrees. At angles above 45 degrees it will pass greater percentages of the light through the glass, but it only approaches 100% transmission at an incident angle of 90°. The panels must be articulated to operate with full efficiency, but this is not done on most roof applications. The result is that they only produce a small fraction of the energy expected of them and attributed to them in all the sales brochures.

It is not difficult to articulate a single panel, but building an array causes them to shade one another unless the array is on a 45 degree angle slope, which rarely happens. The geometry of most locations defeats the panels in terms of presenting them to the sun.

The outputs quoted we have seen have all assumed nine to 12 hours of sunlight exposure, 100% transmission or close to it, but physics says they will only produce full output for two hours and some portion of it for four hours with none for the rest of the day.

The solar boiler systems seen in California deserts are more effective, but they use large areas of land and any birds flying over them get cooked. They have caused the PETA and other bird huggers great distress resulting in many angry letters and threats to the elected folk. They appear not to have much of a future and all have had so many mechanical and maintenance problems they will likely be seen as yet another huge "green" flop.

The bottom line in the energy story is that we cannot beat fossil fuel. With all the gas and liquid petroleum that can be obtained with new drilling techniques we now have an abundance of sources in North America and we should exploit them, but have been constrained by elected people who are afraid of environmentalists and have ruined the education of the scientists who could help them out of the mess they have created had they been well taught.

To date our national Administration has made a political boondoggle of "renewable energy" using it to get huge political contributions from companies that got Federal loan guarantees: "Abound Solar" for $400 million, "A123 Systems" got $279 million, "First Solar" received $646 million and "Solyndra" $535 million. Each gave a few hundred thousand Dollars to the Obama campaign and we have no way of knowing what portion of loan proceeds went into foreign bank number accounts. There are 120 countries doing this kind of banking.

Forbes Magazine has estimated that "global warming," or "climate change" spending and damage to the economy is now costing us $1.75 trillion per year. We know they have destroyed science education and there is no way of adding up that damage, but it is certainly in the trillions.

How can we beat the "global warming," "climate change" people and restore honesty in science? That is our real problem.

CO_2 Is Innocent!

These two bottles are a "demo" experiment to show the effect of adding small amounts of CO_2 to the atmosphere revealing the fraud in what has been claimed since 1988. Over $550 billion was wasted on thousands of "papers" that were a full-employment scam for Ph.D.s in physical sciences. All knew what they were doing and should lose their degrees as they have corrupted science plus science education and cost America over $7 trillion according to analysts at Forbes Magazine, a top financial publication.

Warmist Dr. Joe Romm claims atmospheric CO_2 will rise to 910 parts per million, ppm, by the year 2100 where today it is 390 ppm, a 233% increase in 84 years and turn America into Death Valley. We can simulate this with two

2.5 liter plastic soda bottles for less than $10, without a million Dollar grant, trip to exotic places to read papers, eat truffles and drink Champaign on Federal tax Dollars.

We only need two lab "stick" thermometers, 650 milliliters of pure water, 1/8th tsp. of baking soda and a few drops of Distilled White Vinegar to make the 2100 AD atmosphere of America predicted by Dr. Romm.

The bottles must be clear, not tinted and 2.5 liters. There are other sizes and size is critical. Lab thermometers are $2.39 each on Ebay. Baking soda and "White Distilled Vinegar" are common pantry items, medicine droppers are found in most homes or $1.00 in a drug store.

Thermometer accuracy is easily confirmed by putting them into a tall glass of ice and water, where they should say 0° Celsius. Then, into a pot of boiling water where they will say 100° Celsius if you are at sea level, along with 80% of the world population. If you are higher the boiling point will be lower.

In Chicago, at 650 ft, it is 99.5°C, which is the level for most of the nation east of the Mississippi, but in Denver, at 5280 ft. it is 95°C and the far west is all over the place with Death Valley below sea level where the boiling point is 101°C! Nonetheless, we are only interested in how the thermometers track one another. They must be consistent. If one shows a degree warmer consistently through the 20° to 35° Celsius range then it is usable. Range is the issue.

Put both thermometers in a glass two-cup measure with one cup of cold tap water. Wait a two minutes then read the thermometers. They will say 15 to 20 degrees. Add one cup of hot water from that tap and track the change as it will rise from the temperature of house cold water to the higher temperature of the hot and cold water mix. And, they should move through the range together or with the same difference consistently. If not return the one showing the incorrect temperature in ice water. That is one point that is the same everywhere.

Plastic bottle caps are drilled by a Phillips screwdriver with a 1/4 inch shaft with the tip held over a candle flame for 30 seconds. The handle insulates heat so you can hold it. The hot tip goes through plastic like butter leaving a hole just large enough for a thermometer shaft. Both thermometers are pushed into the caps three inches while the plastic is soft. On cooling they freeze in place, but can be removed later with careful twisting using your thumb and forefinger near the cap.

The volume of the "2.5 liter" soda bottle is actually 2.725 liters. Use bottled water, to avoid municipal water chlorine and fluorine, putting 325 milliliters into each for a net air volume of 2.4 liters over water. This simulates Earth's air as 71% is covered by water and the green areas put almost as much water vapor into air as do the seas.

2.4 liters is 1/10th "molar volume" of air at 20° Celsius, a common room temperature in the United States. "Molar" refers to "mole," a contraction of "molecular" and means the volume of a gas with a mass of one molecular weight in grams. To determine how much baking soda and vinegar to use in creating Dr. Romm's feared atmosphere of 2100 AD we use basic chemistry. It is all based on the relative weights of the elements. Hydrogen, H, the lightest element, is defined as weighing one atomic weight unit. It has two atoms in each molecule so it has a "mole" weight of two grams and a volume of 24 liters at 20 Celsius degrees. The same is true for 32 grams of oxygen, O_2. 28 grams of nitrogen, N_2, or 44 grams of CO_2, carbon dioxide. Every one has a molecular volume of 24 liters at 20°C and each contain the same number of molecules.

Air is a mixture of three gases and eight "trace" gases, which means we know they are present, but they are of no consequence. CO_2 is in that class having only 0.04%. To be a significant part of the atmosphere it must have more than one percent. CO_2 has been politicized for money and power. Politics runs on other people's money through fear and intimidation. The atmosphere has become the newest playground for politicians looking for new taxes, controls and bureaus to build. The fact that air is free and they have not been able to control or tax it drives them nuts.

The only gas changing quantity in air is water vapor as it can exist as a solid, liquid or gas in temperature ranges on Earth. When it condenses the volume shrinks by a factor of 1333. No other atmospheric gas does that. It absorbs energy from sunlight. No other gas in air does it as well and CO_2 only one-seventh as well. There is so little CO_2 it was ignored until Jim Hansen created "global warming," "forcing," "greenhouse gases" and other science myths for money. He found it easy as science is based on myths and scientists are good at inventing fables which may surprise those who are unfamiliar with the ways of science or what is involved in defining an hypothesis and developing a new theory.

No physical scientist believes atoms are little solar systems with electrons like planets and a "sun-like" nucleus. It is a concept we can use, teach and manipulate mathematically, but it is a myth. Matter at the atomic level is much more complex than taught. We experience three dimensions plus time as dimensions mathematically, but modern physics says there are seven more. In this exercise we will only deal with the three of the "x,y,z" axes and time.

In our "Atmosphere of 2100 AD" simulator we will add 520 ppm, parts per million, of CO_2 to the present day 390 ppm with baking soda and White Distilled Vinegar, which contains acetic acid, using the reaction:

$NaHCO_3 + CH_3COOH \text{ ---> } CH_3COONa + H_2O + CO_2$

To simulate the atmosphere of 2100 AD feared by Dr. Joe Romm and others who claim it will have 910 ppm of CO_2 We put sodium bicarbonate and acetic acid in the bottle to make the additional 520 ppm we need to have 910 ppm CO_2. The reaction for this is our guide and key to using the correct amounts of baking soda and White Distilled Vinegar to make precisely what we need.

Sodium bicarbonate, $NaHCO_3$, acetic acid, CH_3COOH combine to make sodium acetate, water and carbon dioxide gas. The molecular weights are 84 grams/mole for sodium bicarbonate, 60 grams per mole for acetic acid to produce one mole of CO_2, 44g or 24,000 ml of gas at 20°C, but we only want a tiny amount and our problem is precision.

2016 air has 0.039% CO_2, 9.36 milliliters per mole or 0.936 ml/0.1 mole in our case as we are working with 1/10th molar volume. For the atmosphere of 2100 AD with 910 ppm CO_2 we need 910/390 as much CO_2 by: 0.936 ml x 910/390 = 2.184 ml, an additional 2.184 - 0.936

= 1.25 ml of CO_2 for our 0.1 mole bottle. This requires only 1.25ml/24000 ml = 0.0000521 mole each of baking soda and White Distilled Vinegar, per the equation.

To find the amount of baking soda we take 72 x 0.0000521 = 0.00375 grams which is an amount we cannot measure accurately as a solid but we could dissolve a gram in a liter of water, which would then have 0.001gram/ml or 0.00005 grams/drop and obtain the 0.0375 grams with 0.000375/ 0.00005 = 7.5 drops which we could achieve by diluting 100% and then counting out 15 drops. Such are the manipulations required in a chemistry lab.

An easier strategy is to control the amount of CO_2 produced with the acetic acid as it is easier to measure 0.0000521 mole of acetic acid where White Distilled Vinegar is only 5% acetic acid. It contains 5 grams per liter which is 50g/60 g/mole = 0.833 mole/liter or 0.000833 moles/ml or 0.000833 mole/ml/20 drop/ml = 0.0000417 mole/drop.

0.0000521 mole of acid would be 0.0000521/0.0000417 = 1.25 drops and we can get very close to a perfect dose by taking two drops of the acid, adding eight drops of water and then using six drops of that in the simulator for five sixths of 1.25 or 1.04 drops of the acid. That is about as close to perfect as we can get with simple equipment and procedures.

Let it sit overnight to be sure all of the acid has reacted. Then put it in the sunlight next to the 2016 air bottle and see the 2100 AD air fail to rise in temperature any faster than the 2016 bottle. Dr. Joe Romm is wrong.

This can be done outside, but we favor a south facing window sill as the bottles are easily tipped by an errant breeze and window glass does not block infrared, IR, energy. When the sun is low little IR comes through all the air between us. After 10 AM the sun angle is above 45 degrees and IR passes through the glass where it cannot when the angle is less than 45 degrees. The thermometers are shaded from direct sun with foil "hats" and are reading the temperature of the air in the flasks and not the direct sunlight.

To test the effect of increasing CO_2 in our atmosphere from the 390 ppm of today to 910 ppm for the year 2100 AD as claimed by Dr. Joe Romm, we only need to put half a 1/4 teaspoon measure of baking soda in the "2100 AD" bottle, swirling to dissolve it. Then put one drop of Distilled White Vinegar in a small glass or vessel, add six drops of water and then put five drops of that solution in the simulator to produce the Romm feared air of 2100 AD.

Restore the cap with the thermometer and let it stand overnight to react fully. Put the bottles in a window sill and record their temperatures each hour.

According to Dr. Romm the 2100 AD bottle will get hot, maybe go over 40 Celsius degrees, 104 Fahrenheit! But what happens? The 2100 AD bottle tracks precisely with the 2016 bottle air with less CO_2 in it! CO_2 was an insignificant component of the atmosphere at 390 ppm and it still is at 910 ppm. It is just that simple.

As we have seen, "warmists" overplay everything from polar bear deaths to all the ice melting at the poles when there are more polar bears than ever in history, probably because they are eating well from all the garbage left by warmist scientists. At this writing, Dr. Romm had to be rescued when his chartered ship became ice-bound in the Arctic. He was there to show that all the ice had gone!

If Dr. Romm were correct the air in the "2100 AD" bottle would heat faster and greater than the "2016 AD" flask and go above 40° C by noon, but it fails to. This puts his, and others alarmist claims into denial. We realize saying the sun will not come up on January 1, 2100 AD sells many more books and get more grants than saying it will. Such is the avarice that drives many Ph.D. scientists to commit professional fraud. Where this is so easily determined they should lose their degrees and posts and go to prison. This alone destroys the anthropogenic global warming concept. Let us carry on...

What if we put enough CO_2 in the atmosphere for it to be considered by professional meteorologists? One percent? That requires we generate 24 ml of CO_2 in the bottle. Where one mole each of baking soda and vinegar make 24,000 milliliters of CO_2 we need only 1/1000th mole to make 24 milliliters. That would be: 0.084 grams of baking soda and 0.060 grams of acetic acid, the amount in 0.060g/0.050g/ml = 1.2 ml or 24 drops. Again, we overcharge baking soda and limit the reaction with measured White Distilled Vinegar and chart the results.

The window in which the bottles stand faces south with a ten degree deviation to the east and were the trial was done virtually on the day of the solstice, change of seasons, it rose due east. The room was at 20°C so there was heating for a bit over an hour raising the temperature in the flasks about four Celsius degrees. The day was cloudless.

Two Atmospheres Experiment			09.20.16
20 C room			
Time	.04%CO2	1%CO2	change
8	24.5	24.2	-0.3
9	31	30.5	-0.05
10	34.8	33.8	-1
11	37	36	-1
12	38	37	-1
1	39	38	-1
2	38	37	-1
3	36	35	-1
4	33	32	-1
5	33	32	-1

Instead of capturing heat and turning the bottle into a pizza oven the 10,000 ppm CO_2 air has forced water vapor out of the air and into the water, per Le Chatelier, and the temperature has declined. And, this is with a 25.6 times increase in the CO_2. No animal or person would be harmed, the plants would love it and we would save much of our water, 80% of which goes to food plants to capture what little CO_2 we currently have! We need more CO_2!

Critical to understanding this outcome is that water vapor is seven times the catcher of IR energy from sunlight as carbon dioxide, CO_2, and heating our atmosphere is thus controlled by the Le Chatelier Principle which states: "When a system at equilibrium is subjected to a change of pressure, volume, concentration or temperature it reacts to counteract that change and establish a new equilibrium." Where air only has two players in the heating case it is expressed as:

$$\frac{[H_2Og] \times [CO_2g]}{[H_2Ol]} = Kt$$

The "[" brackets mean "moles/liter," "g" is for gas or vapor, "l" is for liquid, "K" means "Constant" and "t" is temperature in degrees Kelvin. Water is the only participant that can change from gas to liquid; CO_2 cannot and we can solve for water vapor to determine why the system behaves as it does.

Using the values of 0.03 moles of H_2O water vapor, 0.00039 mole and 55.5 mole/liter for liquid water, Kt = 2.1 x 10^-7. If we revise the equation to solve for water vapor:

$$[H_2Og] = \frac{[H_2Ol] \times Kt}{[CO_2]}$$

Then, use the values to calculate the molarity of water vapor for 2016 air with 390 ppm CO_2 we get 0.03 mole and for having 1% CO_2 or 0.00117 mole H_2O. This is only 3.9% of the water vapor we had at 390 ppm CO_2 as it has driven out 96.1% of the water vapor. Where water vapor is seven times the heater as CO_2 the heating capacity of the air has been reduced to 96.1% x 1/7 = 13.7% of normal solar heating effect. Carbon dioxide is not a heat booster in the atmosphere it is a cooling agent!

Jim Hansen and friends had it right in 1971 when they made a presentation to Congress presenting the hypothesis we could be entering an man-driven ice age. Where they did not have a plan for taxing, controlling, building bureaus to bring us to our knees Congress thought it was one big snore. So Jim returned to his desk at NASA to fester for 17 years until Albert Gore, Jr. showed up.

Vostok Ice Core, and other, evidence had the physical

science community talking of a coming ice age for decades before Jim Hansen's ravings about "CO_2 forcing," a phenomenon not found in any Physics textbook before Dr. Hansen nor has it been explained by him, or anyone, in a paper for 28 years. Hansen doubled-down claiming a "greenhouse" model with CO_2 forming a layer at the top of the troposphere trapping infrared energy in the manner of a greenhouse, hatching the "greenhouse gas" expression and myth.

A "greenhouse" has a clear roof of transparent glass that acts as a one-way window only totally open when the incident light beam is at "right angles," 90 degrees. Below 45 degrees all radiation bounces back to space as the surfaces at that angle that reflect. CO_2 cannot make a "greenhouse" as gases cannot form solids or liquids per physics to be reflectors. "Greenhouse gas" is a lie, but more importantly, a "comity code."

"Comity code" is a phrase uttered by Ph.D.s when making presentations to a general public audience including other Ph.D.s. When he says "Greenhouse Effect" the message to his fellows is, "Let me get away with my fable and I will help you to get away with yours."

The "anthropogenic," man-caused, "global warming" has been the greatest science conspiracy of all time. It has cost America over one trillion Dollars according to the analysts at Forbes Magazine and probably several times that for other countries of the world that have looked to the us for leadership only to be defrauded. Where no great empire has lasted more than 250 years and we are 240 years old will this be the end of America? We can only hope that something as simple as our two bottles can save us.

CO_2 Talking Points

CO_2 is a “trace gas” in air and is insignificant by definition. It would have to be increased by a factor of 2500 to be considered "significant" or "notable." To give it the great power claimed is a crime against physical science.

CO_2 absorbs 1/7th as much IR, heat energy, from sunlight per molecule as water vapor which has 188 times as many molecules capturing 1200 times as much heat producing 99.9% of all “global warming.” CO_2 does only 0.1% of it. Pushing panic about any effect CO_2 could have is clearly a fraud.

There is no “greenhouse effect” in an atmosphere. A greenhouse has a solid, clear cover trapping heat. The atmosphere does not trap heat as gas molecules cannot form surfaces to work as greenhouses that admit and reflect energy depending on sun angle. Gases do not form surfaces as their molecules are not in contact.

The Medieval Warming from 800 AD to 1300 AD Micheal Mann erased for his “hockey stick” was several Fahrenheit degrees warmer than anything “global warmers” fear. It was 500 years of world peace and abundance, longest ever.

Vostock Ice Core data analysis show CO_2 rises followed temperature by 800 years 19 times in 450,000 years. Therefore temperature change is cause and CO_2 change is effect. This alone refutes the anthropogenic global warming hypothesis.

Methane is called "a greenhouse gas 20 to 500 times more potent than CO_2," by Heidi Cullen and Jim Hansen, but it is not per the energy absorption chart at the American Meteorological Society. It has an absorption profile very similar to nitrogen which is classified "transparent" to IR, heat waves and is only present to 18 ppm. "Vegans" blame methane in cow flatulence for global warming in their war against meat consumption.

Carbon combustion generates 80% of our energy. Control and taxing of carbon would give the elected ruling class more power and money than anything since the Magna Carta of 1215 AD.

Most scientists and science educators work for tax supported institutions. They are eager to help government raise more money for them and they love being seen as "saving the planet."

Conclusions

The idea of anthropogenic, man-caused, global warming began as Roger Revelle's scheme for fame and fortune attracting more scoundrels along the way.

Late in life he had a crisis of conscience seeing where his myth was taking America, but the evil genies were out of his lamp. Now it is our hope to put them back to where they belong because anthropogenic, man-caused global warming or "climate change" is the greatest science fraud since the early, pre-scientists claimed to turn lead to gold with the yolks of mythical bird eggs. They also sold their souls to the rulers of the time and occasionally lost their heads for their failures.

While we do not recommend beheading scoundrels in the sciences we do think it would be very appropriate to take away their degrees and prevent them from teaching or working as scientists. By its nature science is work that requires high integrity.

The Revelle Science Prize is awarded by Scripps Institute, the home of the great fraud thanks to the four horsemen of the science apocalypse, Roger Revelle, Charles Keeling, Naomi Oreskes and Albert Gore, Junior.

The Revelle Science Prize has a history of being awarded to the leading skunk in the anthropogenic global warming fraud for money and power. It will one day be seen as a list of the names in science infamy.

It is clear Roger Revelle had a turn of mind by 1988 and lived until 1991. He had the time and clout to publish a book revealing the truth of the matter, but he chose not to. For that we can never forgive him for creating America's vapor tiger.

The End

For Books by Adrian Vance

Search Amazon.com for "Adrian Vance"

See: "The Two Minute Conservative"

At: http://adrianvance.blogspot.com

www.ingramcontent.com/pod-product-compliance
Lightning Source LLC
Chambersburg PA
CBHW051730170526
45167CB00002B/877
9781502596321